HUMANS OF CLIMATE CHANGE

A Cultural Journey
to Explore Climate-Change
Impacts, Solutions, and Hope

KADEN HOGAN

For any enquiries, please send an email to Kaden Hogan at kh@humansofclimatechange.org

ISBN 978-1-7399080-4-1 Paperback
ISBN 978-1-7399080-0-3 Hardback

SUSTAINABILITY CHALLENGES

If you are interested in participating in sustainability challenges that will help you take small, local actions against climate change, please join our "Humans of Climate Change" community.

You will also have a chance to **win money** by taking part in these challenges!

Follow this link: https://community.humansofclimatechange.org or scan the QR code below for further instruction.

JOIN THE COMMUNITY

TABLE OF CONTENTS

INTRODUCTION

Y OU CAN'T HAVE AVOIDED HEARING about climate change. It's in the news all the time – major intergovernmental treaties, reducing carbon use, grants for insulating your home, wildfires, floods, rising sea levels, Greta Thunberg, and the School Strike for Climate.

But it's such a big subject and often presented in quite threatening terms, reminiscent of a hell-and-damnation preacher, that it can be difficult to get your head around it. Statistics get thrown around like confetti, and different deadlines – 2030, or 2050, or the end of the century – what's going on behind all that? It's often easier to switch off or to conclude there's nothing you can do about it, and just give up.

I can sympathize with that! I have a scientific and engineering background. I've read, studied, and worked in industry areas related to climate change and sustainability. And I have a real passion for the environment.

But even with all this, I find it hard to get my head around exactly what a two-degree rise in the Earth's temperature, or a one-foot rise in sea level, could mean. The way many people – the media as well as academics – talk about climate change, is just numbers. However, when I saw a news story about how some Pacific nations could actually disappear under the sea by the end of this century, that made a huge impact on me. That's one of the reasons I wrote this book.

Another is that I've seen how climate change is already begin-ning to make a big impact. It isn't something that *might* happen by the end of the century. It's something that's happening *right now*, and people are having to change the way they live because of it. Some of these changes are small – growing different crops, paint-ing roofs white – and some of these changes are huge – people moving hundreds of miles to find somewhere they can live, and whole cities being evacuated because of the risk of forest fires.

In every culture around the world, storytelling is a big part of how we pass on our traditions and knowledge. And it was a story that brought home to me what the abstract and impersonal term "climate change" really means. This book is a collection of stories about how people all over the world are being affected by chang-es in the environment – and what they're doing to adapt.

This isn't a top-level book about greenhouse gases or long-term forecasts, or to advocate for governments to implement car-bon tax policies. There are plenty of books out there that will take you through the facts and figures. In fact, unless you're a research scientist yourself, some of the science is difficult to grasp in terms of its everyday impact. I believe most of us can make more sense of the science when we're able to track how it affects people's lives in particular places.

So, instead of being a facts-and-figures book or a theoretical text, this book is a trip around the world, from the crystalline soli-tude of the Arctic icefields to the lush green rice paddies of the Me-kong Delta; from the amazing engineering marvels of the Dutch sea defenses to the centuries-old French tradition of winemaking.

Everywhere, individuals are adapting to climate change. It's not governments and politicians and rich philanthropists I'm go-ing to tell you about – it's ordinary people who are helping their environments and communities to adapt. Sometimes, it's the least powerful people who are making the biggest changes – Indigenous peoples like Inuit and the Amazonian Indians, women in patriarchal societies, and young people. They're not waiting for a guy in a suit to do it, or for permission from the authorities, their husbands, or their parents – they're just doing it.

I've been hugely impressed by how adapting to climate change has empowered many of these people. Clearly, no single person holds all the keys to stopping global warming or holding back the rise in sea level. But instead of giving up, individuals and communities all over the world are discovering small actions they can take to have a positive impact – and they're taking them.

In fact, as I researched the stories in this book, I found that these people aren't just fighting for the planet, or even their small part of it. There's much more at stake . . .

But we have a long journey before we get there. So, let's head off to our first destination. And since I've just mentioned global warming, what better place to start than the chilly wastes of the Arctic?

PART I

WARMING TEMPERATURES

1

THE ARCTIC:
THE WAY OF LIFE

T HE HUNTER HEARS ONLY HIS feet on the ice. He is alone, under the big sky, in the silence. Wherever he looks, there is white. He is a dark dot on the landscape, moving slowly, carefully, constantly.

Someone who had not been born here would see no landmarks. But the hunter knows this landscape as he knows the wrinkles in the palms of his own hands. He knows how far he can safely go before the sea ice starts to give way under his weight; he knows how to read the sky and the wind for approaching storms.

He's out for bear. Polar bear.

* * * * *

Inuit of Canada live in what most of us would consider one of the most inhospitable climates in the world. They are hunters and fishers who know how to be self-sufficient in a harsh climate. They are attached to their land and their traditions, so much so that it's difficult to know where the land ends and the traditions begin.

During the short summer, they make jerky from the caribou they hunt; in the winter, they dig holes in frozen lakes to drag up nets full of wriggling char, which can be made into soup or just eaten raw, like sashimi. Hunters head out to fish for salmon, or to hunt seals, walrus, polar bears, and caribou.

In the winter, you have to keep moving – a tent will turn the ice beneath to slush after a while, so moving the tents around is a frequent task in the hunting camp. And if you're walking, you have to keep moving; if you stop, you will freeze; if you fall asleep, you will never wake.

Inuit are moving all year, traditionally living a semi-nomadic life – in snow-built dwellings on the sea ice in winter, hunting seals and fish, then in springtime living on the coast. As summer comes, they move inland to hunt caribou and supplement their diet with berries and birds' eggs. They use the kayak or *umiak* (an Inuit invention) for hunting seals, and bigger boats to look for beluga and bowhead whales.

At least, that's their traditional way of life.

In the old days, even the more permanent villages would split up if the population became too big to be self-sufficient. Inuit didn't define themselves by whom they lived with or next to, but from webs of relationships stemming from adoption, hunting partnerships, or namesakes (children named after a grandfather, or after father's hunting partner, for instance).

Of course, life was hard. Some hunters drowned. Some were lost in blizzards or mauled by bears. There were famines some years. What can you say when you learn that minus 4°F (minus 20°C) is regarded as "not really cold"?

But Inuit had the skills to survive. For instance, they knew the migration patterns of the caribou, which moved north in the spring and south in the fall, across the tundra – a vast, open plain in summer and a frosty wilderness in winter. Every year, caribou would take the same track, only to find Inuit hunters waiting for them. In fact, some Inuit say that Amundsen was such a successful explorer because he learned local skills – how to make the clothing, hunt, use sled dogs, and stay alive on the land. (Amundsen, like Scott of the Antarctic, made it to the South Pole – unlike Scott, he also made it back home.)

Inuit shared everything within the village; they had no money economy. If a hunter had good luck, the food would be shared with all the families. The bare landscape wasn't bare to their eyes. It was full of information and food if you looked hard enough, and to be hunting in it was not loneliness – not to them.

They were a mild and gentle people; anger was seen as by far the worst vice. Instead, they valued patience; the patience of a hunter waiting for his prey, the patience of a family waiting out the winter playing cat's cradle and telling stories. *Ajurnamat*, they'd say of the winter, of hunger, of accidents – "It can't be helped."

But now the ice is melting . . . and things are changing for Inuit.

The new Cold War?

As the northern ice melts, Russia and America have noticed. So has China. New shipping lanes and new mineral resources are becoming available as the ice withdraws.

The Arctic used to be a frozen wasteland, no use to anybody but Inuit. Now it is the "forefront of opportunities and abundance," as U.S. Secretary of State Mike Pompeo said in the Arctic Council meeting in 2019. It's no longer seen as a home, but an "arena" – perhaps for a new Cold War.

Resources include gold, rare earth metals, diamonds, oil, and gas, as well as fisheries. Meanwhile, new shipping lanes reduce the time at sea from Asia to Europe by up to two weeks by taking the Northwest passage up from the Canadian coast and through the Bering Sea to the west coast of the U.S. and the Pacific, or the Northern Sea Route along the northern coast of Russia and emerging through the same gap at the Bering Sea.

Russia has expanded its Arctic military resource; China is claiming status as a "near-Arctic state." With a commodity-driven economy, Russia is clearly after the natural resources, as is oil-rich Norway, but China appears to see the Arctic as an extension of its claims in the South China Sea.

Strangely, at the 2019 Arctic Council meeting where Mike Pompeo announced a new focus on the Arctic, no mention was made of Indigenous peoples, even though the Council has six permanent participants representing them. Almost no mention was made of wildlife. The focus, instead, was on territorial claims and potential military conflict.

The U.S. has an annual military exercise called "Arctic Edge", and the Canadians have "Operation Nanook", while 50 radar surveillance stations guard the northern frontiers of Canada and the US. Meanwhile, Russia has been investing in new military bases and recently sent three nuclear submarines to surface through the Arctic ice.

The "gold rush" for new resources, like previous plans to exploit the area, has bypassed the Indigenous people. In the nineteenth century, whaling changed the lives of Inuit living in Nunavut, the northernmost territory of Canada, but it contributed little to their welfare – all the wealth flowed south. There are few jobs in the extreme north, and with most Inuit living in permanent settlements now, the old way of life has been destroyed with nothing to replace it.

* * * * *

The melting of the sea ice and other environmental changes have had a massive impact on Inuit. The loss of sea ice makes it far more dangerous for hunters to travel. You can't trust the ice anymore – even seasoned hunters are falling through, misjudging conditions they no longer know how to read with certainty.

Storms have become unpredictable. Hunters can't read the clouds or the wind anymore; the messages are jumbled up. Sometimes they're stranded in a whiteout, unable to move. A good hunter knows how to dig into the snow and quite literally "hole up" while the storm blows over, but even so, some never come home.

Meanwhile, animals have lost much of their habitat. Some have changed their migration patterns so that hunters can't count on the caribou being in the right place at the right time. Other species are now protected by law due to decreasing numbers, but this makes it difficult for hunters and their families to be self-sufficient.

"Country food", as Inuit call their traditionally hunted and foraged food, is a communal resource freely shared out; on the other hand, grocery stores in Nunavut have prices two or three times higher than in the rest of Canada. And synthetic textiles you can buy in the shops aren't as warm as traditional caribou hide and polar bear fur (the words "anorak" and "parka" both come from the Inuktitut language).

Many Inuit have been forced to relocate to permanent settlements, but a bright-yellow cabin in a grid of streets is no replacement for the feeling of freedom out on the ice under the crystalline sky. They have been taken from their true home, their place in the world. So it's perhaps not surprising that Inuit have the highest rate of suicide and depression in the world.

In 2013, Nunavut's then-chief coroner, Padma Suramala, pleaded for Canada's health authorities to help after dealing with 14 suicides – including a 13-year-old – in just two months. The suicide rate is 10 times the Canadian average, and for boys aged between 15 and 19, it's a shocking 40 times higher. Worse, suicide has become normalized in Canadian Inuit society (Schreiber, 2018). (Greenland, another Inuit society, has similar problems.) Everyone knows someone who took their own life.

Drug addiction, alcoholism, and child abuse are all common. This is an unexpected side to climate change – its impact on mental health. To Inuit, it's not just about the loss of their land and their traditions, it's the loss of their identity as a First People.

Some young Inuit boys haven't been out hunting or kayaking since their grandfathers or elderly uncles died. They don't have the skills they need to live off the land, and if there are any jobs available, they're under the bright fluorescent lights of a supermarket, not out amid the pure whiteness of the ice. These young men are restless, longing for a connection to the land but unable to find it.

~ *Based on photo by Camilla Andersen* ~

Some Inuit are using poetry to express their feelings about the distance between the past and the lives they live now. Alootook Ipellie writes about seeing a photograph of boys playing, and in the photograph, there are dog teams going away:

But I didn't know where they were going,

For it was only a photo.

I thought to myself that they were probably going hunting,

To where they would surely find some seals basking on the ice.

Seeing these things made me feel good inside,

And I was happy that I could still see the hidden beauty of the land,

And know the feeling of silence.

A precarious balance under threat

A global rise in temperatures of 2°F (1.2°C) since 1880 doesn't sound like much. You might shrug it off. If you live in a temperate climate, it won't matter that much to you – maybe your skiing holiday will be a bit short on snow; maybe there will be an occasional heat wave.

But the rise has been far more extreme in polar regions. In some areas, the temperature has increased by 45°F (25°C) at times (Samenow, 2018). That is an incredible rise.

And the new warmth has been nibbling away at the ice cover, faster and faster. In fact, it's not nibbling anymore; it's gobbling up the ice. According to NASA, 21,000 square miles (54,000 square kilometers) of ice is being lost every year; that's enough to swallow a country the size of Costa Rica or Switzerland.

The 2014 National Climate Assessment reported that the Arctic Ocean could be completely ice-free in summer before 2050. Add to this the Colorado National Snow and Ice Data Center's opinion that an already steep rate of decline has accelerated in the last decade, and it's likely that, during our lifetime, the entire Arctic will be ice-free during the summer.

The Arctic is a terrifying place. It doesn't obey the rules of the normal world. It's dark for half the year, then has a brief summer of twenty-four-hour days without a sunset. Some of what you think is a solid island is actually made up of ice floating over deep, dark water.

In winter, the ice grows, and blizzards sweep the land; in summer, the ice melts, drop by drop.

This has been the eternal pattern of the seasons. It's a delicate balance. But now, the balance has tipped, and the summer melts more than the winter can ever replace.

2007 saw the most significant Arctic ice loss ever, unexpected even by the pessimists (Shea, 2019). It seems that a new ocean is being born. Who's going to fish it? There are no rules for this new sea. Chinese super trawlers could strip it of fish within years.

And this is not just a problem for Inuit. It's a problem for the entire world. The Arctic ice reflects 80 percent of sunlight back into space, keeping the planet cool. But water absorbs 90 percent of sunlight – which heats the oceans. And it has become a vicious spiral.

We need to do something . . .

* * * * *

There is a silver lining to this cloud. Opening up the new sea lanes could help cut marine pollution and the carbon footprint of world trade.

The Third IMO GHG Study in 2014 estimated that shipping emitted just over 3 percent of total global carbon dioxide. A further update to the study predicted that this could increase to 10 percent if other sectors decarbonized successfully and shipping wasn't able to do so.

Here's where the new sea routes could really help. To currently get cargo from Shanghai to Rotterdam, a ship has to go south of India, then up the Red Sea and the Suez Canal – which as we saw recently with the grounding of the Ever Given is a major bottleneck – then through the Mediterranean and up the Atlantic coast of Europe to the Netherlands. Using the northern route could cut that distance in half.

So that would halve the carbon footprint of every iPhone exported from China to Europe, for instance.

* * * * *

Back in Nunavut a boy is walking.

He's walking down the road. Mud and gravel.

He reaches the end. He walks back.

There's nothing to do here. He can go home and play computer games. He can go home and do nothing. Or he can walk.

Back past the shipping container. Back past the bright-red house. Back into town.

And then back out again.

Walking, walking, because he doesn't know what else to do.

Turning back at the end of the road because he doesn't know how to find his way in the snowscape. He wants to be in that white silence but he's afraid of it.

Walking, walking, because he doesn't want to have to listen to the voice in his head that says, *Boy, you're lost.*

He's staring at the river. He wonders how long he'd last in the tumble of icy water if he jumped.

He walks on.

* * * * *

And meanwhile, what of Inuit as a whole? For an Inuk man, life is hunting and hunting is life. But because of compulsory residential education programs that took the last generation away from their land and culture, and the aging of the last generation who genuinely lived on and from the land, young Inuit have lost that connection.

Inuit culture is based around sharing, but it's also based on self-reliance, silence, and strength. That makes it hard for Inuit youngsters to seek help; asking for help is seen as a weakness. A standard American-style mental health program might not get many takers.

Instead, local community programs aim to reconnect the youth to the land, by offering hunting and camping trips with the elders, or "country-food" community meals.

Ilisaqsivik Society in the little town of Clyde River was founded in 1997. It's a grassroots effort, aiming to find a wellness program based on Inuit values, not imported from elsewhere. Its counselors speak Inuktitut, and part of the mental health program is based around hunting trips, where elders help younger Inuit understand their land and the hunting culture.

Ilisaqsivik and similar programs don't just help individuals; they also help bring the community back together. Intergenerational trauma is being tackled by creating mentorship and giving younger Inuit older friends – not counselors or advisers, but friends they can talk to.

Other Inuit reconnect with their culture through art and poetry. Art is now a big contributor to the local economy; in 2015, Inuit art brought CAN$87m into Canada. This doesn't just provide Inuit communities with an income, it also allows artists to stay in touch with their roots and work with traditional materials in both handed-down and contemporary ways.

"Country food" – hunger for Inuit roots

Another grassroots institution, Qajuqturvik Food Centre in Iqaluit, the capital of Nunavut, is reintroducing "country food" to Inuit diet. Iqaluit (population 8,000) means "place of many fish", so it's an appropriate location for this initiative!

The food center is particularly important since the changing migration patterns of animals, together with Inuit's newly settled lifestyle, have brought high food insecurity. 70 percent of Inuit children live in households where it's a problem. Supermarkets charge high prices for imported food from elsewhere in Canada – CAN$20 for a kilo of fruit, for instance.

And Hawaiian pizza, hot dogs, and pastries are not traditional Inuit foods. Instead, "country food" includes *muktuk* (whale skin and blubber), seal, muskox, goose, salmon, and Arctic char, with bear as an occasional treat (some people say the paws are the best bit). It's a high-protein, practically paleo diet, with fish often eaten raw and meat dried as jerky.

The food center provides traditional meals by partnering with local hunters and buying their produce. Like Ilisaqsivik, it pairs up young boys with hunters to learn the traditional knowledge, not just hunting but also reading the landscape and identifying plants. Then, of course, there's always a discussion about how the food should be prepared and cooked. It's more than a food bank – it's a cultural center.

Tourism is becoming a draw, and with more open water, cruise ships can now bring passengers to Nunavut in an expanded season – though that can cause pollution, and cruise ships don't always contribute to the local economy.

One Inuit community, the Qikiqtani Inuit Association, has introduced fees of CAN$150 per plane landing and CAN$100 per helicopter landing, with a CAN$50-a-day camping fee (Deuling, 2019). Many local tour operators support the fees; they already organize homestays in local villages and use Indigenous boat captains and guides. They also have a strict application process for tour operators to ensure that they are aware of how to minimize their environmental footprint.

While the geopolitical future of the Arctic remains uncertain, Inuit are showing that they can adapt to what could have been disastrous trends by reclaiming their traditional culture.

Maybe we should be listening to Inuit?

2

BURGUNDY, FRANCE:
Changing Climate, Changing Tastes

D ING, DANG, DONG – THE bells of the village church in
the valley are ringing out. You can see its gray spire amid
the little stone houses surrounding it. But up on the hill here, it's
a crisp morning in the vineyards, the leaves just coming, right at
the start of the season.

On every south-facing slope, rows of trimmed vines line up
neatly. Some of them are twisted, their bark knurled, like wizened
ancients; others are more recently planted, straight and smart like
model soldiers. At the moment, you can still see the crumbly soil
between them; later, the growing vine leaves will hide it little by
little until all you can see is green, green everywhere.

This is Burgundy, the very heart of France. Around here, it's
not grand chateaux but small local wine producers that own most
of the vineyards; there are over 5,000 of them, and some families
have owned their lands since the 17th century. Here, to be a wine-
maker isn't just a job; it's a vocation and an inheritance.

To understand Burgundy, you must understand *terroir*, pronounced "tare-waa" (unless you can manage the authentic French "r" at the back of the throat). *Terroir* is a value that's crucial to French identity. What is it, exactly? It's the character and taste of a local area, the geology, the landscape, the exact weather patterns, and the local traditions that create a particular food or wine.

It's a very personal, very small-scale value. If you think of a huge prairie full of big fields of corn . . . *terroir* is exactly the opposite of that.

Let's walk into the village. Outside the boulangerie, there's already a queue – people are buying their daily bread, but because it's Sunday, they're also buying an *éclair au café* or a *tarte au citron* for a special after-lunch treat.

And the food will be good. Burgundy is renowned for its filling food: ham with parsley sauce, *bœuf bourguignon* (a rich beef stew), *coq au vin* (chicken braised with mushrooms) . . . and the not-so-secret ingredient in both the last two is, of course, Burgundy red wine.

But though the village still looks much as it might have done in a sepia postcard from the 1900s, the wine, somehow, doesn't taste quite the same anymore.

The formula: soil + slope + sun + grapes = wine

Wine isn't just about grape varieties (the main varieties here in Burgundy are pinot noir for red wine and chardonnay for white). It's also about the soil. (One reason that some southern English winemakers, like Nyetimber, are doing so well is because they're on the same chalk belt as the Champagne country. And they're also benefiting from the warmer summers.) Here, the soil is a Kimmeridgian limestone, rich in calcium, made of up millions of minuscule sea creatures' fossils.

Wine is also about the exact amount of sun the grapes receive. That depends on the weather, but it also depends on the slope of the hill and the exact orientation of the vine rows. It's also about the exact length of the growing season; how many days of sunlight in total the grapes get before they're picked. And it's also about the individual wine grower's skills; harvesting the grapes at just the right time, neither too early nor too late.

These things can come down to a very tiny area indeed. When we talk about Burgundy, we're referring to several different regions – Chablis, Côte-d'Or, Mâconnais, Beaujolais, Côte de Beaune, and then there are different *appellations* within each region, like Meursault, Puligny-Montrachet, Chassagne-Montrachet or Saint-Aubain. Moreover, within each of those, individual vineyards may be renowned, while others are of lesser quality.

For instance, in the northern area – the Côte-d'Or – the white wines are floral, aromatic, light, and crisp; sometimes, they almost taste like apples. Aligoté, from the Côte Chalonnaise just a little further south, has citrus, flinty, honey notes accompanying that floral aroma. And in the two villages of Rully and Chagny, they make sparkling wine – Crémant de Bourgogne. It's amazing how different wines from vineyards just 10 or 20 kilometers apart can be.

Choosing the right time to harvest is exceptionally important. If it's too late, the sugar content in the grapes will be too high. Pick too early, though, and there won't be enough acidity to give it a refreshing taste; nor enough tannin or phenolic content to provide the richness and complexity you want in the flavor.

The grape harvest usually begins in September. But in 2020, the hot summer got the season kickstarted by August 12th, and it was all over by mid-September; that's one of the earliest harvests anyone can remember.

But we don't have to rely on winemakers' memories. We have logs of the harvest dates going back to 1354 (Labbé *et al.*, 2019) – when the cathedral of Notre-Dame de Paris had just been given its final touches. We even have two separate series for much of that time, some from Dijon, and a complete run from Beaune. So we know just when grapes were harvested – and by looking at that series of dates, you also have a good idea how hot it was that year.

* * * * *

Imagine you're in Beaune, the center of the wine country, in midsummer.

It's crazily hot. The air sears your lungs as you breathe.

The bells of Notre-Dame church are ringing. Further off, you can hear the Hôtel-Dieu's three bells, far in the distance, and Saint-Nicolas' four.

And now you hear the procession; the canons are chanting, and there's such a crowd that you need to stand on tiptoe to see the "Black" Madonna holding the Christ Child on her lap, carried on the shoulders of four strong men, and a priest behind in his richly embroidered vestments carrying a gilded, bejeweled reliquary.

It's too hot. You're nearly fainting. And you hope, against all hope, that the Madonna will bring some rain to cool you down, and to cool down the vineyards, where the grapes are already shriveling in the heat.

Beaune depends on its grape harvest for its wine – the best wine in France. And the way things are, in the Year of Our Lord one thousand, five hundred and forty, you don't know whether there will be any grapes this harvest time.

* * * * *

We know about the heat that year from the winemakers' records, and about the procession from the church archives.

But global warming means those old records may not be so helpful anymore, and winemakers' memories of a lifetime of harvest dates represent knowledge that now may be out of date. Some well-regarded older winemakers, with nearly 60 years of experience, say they cannot remember a summer as hot as these.

If you look at a chart of dates up to about the 1980s, with a few variations, harvest dates look pretty similar over time. But the last 30 years have seen the harvest coming on average two weeks earlier. And because the grapes are more sugary by the time they reach the winepress, the alcohol content increased from an average of 12.5 percent in 1971 to 14.8 percent in 2001. In 2018 some reds were over 15 percent alcohol by volume (ABV).

By 2050 the harvest could regularly be in mid-August, an entire month and a half earlier than normal. But the grapes won't start growing any earlier, so they will have a shorter time to mature; inevitably, they'll have a different balance between sugar, acid, and tannin, and the wine they make won't taste the same.

To make a good chardonnay, for instance, you want to end up with a refreshing light wine, without too much sugar or a syrupy mouthfeel. But with warmer summers, the grapes ripen faster and have a much higher sugar content when they're harvested. Then you end up with a heavy, sugary wine – it's like getting a bagel when you expected a baguette.

* * * * *

As well as winemakers, scientists have been looking at the harvest records – the longest phenology records in Europe. (Phenology is the study of the impact of climate and habitat on recurring biological events, like the date that birds build their nests or when and where they migrate.) They've also calibrated these records against other evidence, such as tree rings, which show wider rings in a rainier or warmer season, and against the long Paris temperature series, running from 1659 to 2018, as well as against other documentary evidence.

Up to 1987, the grapes were picked from September 28th onwards. From 1988 onwards, harvests have been on average 13 days earlier. Crazily hot summers like 1540, when the grapes dried on the vines, were outliers in the past – now they have become standard. There have been fluctuations before, but they always came back to a moderate average in the end – until now.

* * * * *

Burgundy isn't mountainous, but it *is* upland, so it has escaped the heat waves that have hit lower areas of the country. And it's fairly far north, unlike the southern wine-growing area around Bordeaux, where climate change is a real threat.

In fact, 2021 had a different problem to throw at the winegrowers: a late frost and not one, but three nights of hard freeze. The traditional method of lighting a candle at the foot of each vine – creating a flickering, living landscape in the night – didn't work, nor did the modern method of blowing hot air from helicopters. The chill was just too severe.

Fifty percent of the vintage is already lost. Chardonnay is the worst hit, as the chardonnay grapes are early starters; the red pinot-noir vine awakens more slowly in spring, so there are hopes more of that harvest can be saved. "It's a disaster, it's a disaster," said one vigneron. The government has now declared the freeze an agricultural disaster. It will compensate vignerons for the loss, but that doesn't make it emotionally any easier to see an entire year's harvest destroyed.

Winters in the region are becoming colder. The French often talk about the "Saints de Glace" – the Ice Saints – St Mamert, St Pancras, and St Servais, whose feasts are celebrated from May 11th to 13th. Their saint days often bring a late frost, but usually, it's not severe enough to harm the vines. More extreme variance in temperatures, however, could be a long-term threat to the more northerly vineyards.

Winemakers innovate for success

Still, Burgundy wines look as if they have a future. Winemakers could adjust by bringing in new grape varieties . . . that's already happened in Bordeaux, where the previously forbidden Marselan and Touriga Nacional grapes are now allowed. But for now, Burgundy only grows pinot noir and gamay for reds, and chardonnay and aligoté for the white wines.

Mâcon, in the south of Burgundy, has already started to change. In the 1970s, most of the growers in this southerly area were selling their grapes to big cooperatives to make table wine. It wasn't considered high enough quality for anything better. But younger growers decided to make their own wines instead of selling to the co-op. It's warmer here, where the harvest begins two full weeks before more northerly Chablis; the most southern vineyards, in the Pouilly-Fuissé *appellation*, make wines with lovely honeysuckle and stone-fruit flavors and even notes of pineapple and peach.

The Mâcon growers started the Association Les Artisans Vignerons de Bourgogne du Sud – "Craft Winemakers of Southern Burgundy" – in 2004, and winemakers like Jacques and Nathalie Saumaize in Vergisson have gained an increasing appreciation for their vintages. The Bret family took its La Soufrandière vineyard back from the co-ops in 1998, and now makes organic, biodynamically produced wines.

Other winemakers have their own innovations. Etienne Grivot tricked his vines by waiting a few weeks before plowing the vineyards. Plowing warms the soil, which in turn wakens the vines from their winter snooze. So by turning off the alarm clock, he let his grapes take their time, and grow more slowly. Many vineyards had already harvested when the rain came at the beginning of September 2018 – but his later grapes were able to benefit from the rain, and he picked them a week later.

Vine leaves used to be trimmed towards the end of the summer so that the grapes could ripen unshaded, in full sun. Again, many winemakers are now deciding to skip this part of the process, or only trim the leaves on the north side of the rows, or those lower down (which lets air circulate and helps avoid mildew).

Organic cultivation, apparently, also helps. If you give vines fertilizer, they get used to "drinking" from the surface; if you don't, they push their roots much further into the soil for nutrients and water.

Winemakers also have to deal with picking grapes when it's hot. Some now cool down the grapes, which have been harvested in the heat of the day, before processing them. Other estates will even stop picking altogether if the temperature gets too high, opting to wait for the cooler evening.

And the pecking order is changing, too. The best quality wines used to be made from grapes grown at the bottom of the slopes – but now, just being 50 meters higher can make a big difference, and the upslope wines are improving markedly. That may hurt some of the best-known names – but it could be a big opportunity for younger, more innovative winemakers.

And for wine buyers? It means that if you're smart, you can buy a wine that missed out on Grand Cru status when the maps were drawn up in the 1861 classification – but in 2021, is a much better wine than the expensive, downhill Grand Cru!

3

MALI AND ETHIOPIA:
FROM ALLY TO ENEMY

YOU SLATHER YOURSELF AND YOUR bedclothes in an anti-mozzie spray. You burn a mosquito coil that makes you cough. You cover yourself up. You take pills that make your eyes hurt when you're in the sun and make you dizzy half the time.

And you do all this because you know just one mosquito bite could give you malaria, and because more than a million people die of malaria every year.

But you feel good because you've been responsible and protected yourself. You're going to get a great night's rest . . .

. . . until, of course, you hear that unmistakable buzzing and think, "Damn, there's one of them in here with me!"

* * * * *

A mosquito is such a tiny yet dangerous creature. And, as they say, the female of the species is deadlier than the male.

A male mosquito is harmless. He won't bite. But the females, on the other hand, need the protein they find in blood to grow their eggs. They can smell sweat and human breath, and zero in on a nice soft patch of skin where they can stab their proboscis and suck the blood that will feed the eggs they're carrying.

But the mosquito's proboscis isn't a one-way street. As the mosquito sucks in blood, she's letting tiny parasites go the other way, entering her victim's bloodstream. Eventually, they find the liver, where they reproduce and cycle back into the bloodstream, killing red blood cells. The cycle continues, destroying more red blood cells every time and reducing the amount of oxygen, glucose, and iron that is able to reach the vital organs. Meanwhile, the victim's experience matches the life cycle of the parasites: chills, fever, and headaches, cycling through every two or three days.

And a female mosquito that *doesn't* have malaria is just as dangerous as one that does. If she bites an infected human, she becomes infectious after a couple of weeks and can pass the disease on to anyone she bites.

It's like a vampire movie crossed with a zombie apocalypse.

Out of a total of 3,500 mosquito species – 850 species in Africa alone – only 40 spread malaria, but they are an extremely efficient way of spreading it (Noone, 2017). Many victims don't even realize they've been bitten while they slept.

If you live in a northern, temperate climate, you probably don't bother about any of this unless you're taking a trip to the tropics. Malaria mosquitoes prefer a hot, humid climate. Their comfort zone is 77°F (25°C) or slightly above, and they need shallow areas of stagnant water where they can lay their eggs – drainage channels, marshes, swamps, puddles, pools, even water tanks. At the moment, that's exactly what Mali, in West Africa, gives a mosquito, particularly in the rainy season, from June to September.

A dozen ways to beat mosquitoes

West Africa has the highest rates of malarial infection in the world. 340 million people are at risk (McSweeney, 2016), and hundreds of thousands die each year. In Mali alone, with a population of 12 million, there are 800,000 recorded cases a year – and this probably understates the problem, as many cases will be unrecorded. Malaria accounts for 17 percent of child deaths in the country.

~ Based on photo by Charles Nambasi ~

Traditional methods of malaria prevention concentrate on vector control – trying to stop the mosquitoes from getting to people or finding places to breed. Standing water is drained, or watercourses dug to make the water flow faster, thus reducing suitable places for breeding. Mosquito nets for beds, and screens for doors and windows, prevent mosquitoes from coming into the house. And, of course, pesticide spraying can get rid of millions of mosquitoes at a time – but it can affect other species too, including humans.

Anti-malarial drugs are another way to beat the disease, though many have adverse side effects and can't be taken for more than six months, or by pregnant women. Worse, mosquitoes are now becoming immune to many of the earlier generations of anti-malarial drugs, and are increasingly developing resistance to insecticides. We have to get better at fighting malaria, or the progress that has already been made could be reversed (Kushner, 2020).

So, more novel approaches are now being examined. For instance, genetic engineering could be used to change the mosquito's ability to breed. Creating "selfish" genetic elements that make male mosquitoes infertile will eventually eradicate the mozzies. A variation of this approach, the sterile insect technique (SIT), has eradicated some nasties – screwworm, Zanzibar tsetse flies, and the melon fly in Okinawa, Japan – but it only works if you've already suppressed the population through broader vector-control techniques.

Researchers of Imperial College London's Target Malaria research consortium believe they're onto a winner; (Noone, 2017) a genetic-engineering technique that targets just the mosquito, with no impact on other species. The concept was already mooted in 2003, but it's only the recent invention of the CRISPR/Cas9 gene-editing tool that's made it possible to do so effectively and at a relatively low cost.

Another idea is to infect male mosquitoes with Wolbachia. This genus of bacteria makes them destroy the eggs of uninfected females after mating. The release of infected male mosquitoes into the wild has already been trialed in Australia and California, as well as in Indonesia where it appears to have been highly successful against dengue fever, another mosquito-transmitted disease.

Vaccines are another possibility. But the RTS,S vaccine developed by PATH and GlaxoSmithKline is only 30 percent to 40 percent effective. Most vaccines are 85 percent to 95 percent effective. Now the University of Oxford has developed a malaria vaccine (R21/Matrix-M) which is starting its phase III clinical trial; it might do better, but it is unlikely to be fully licensed until 2023.

So far, the mosquitoes are still winning. But there is another improbable ally in the fight against malaria – climate change. At least, it's your ally if you live in certain areas of Mali.

* * * * *

Mosquitoes are climate sensitive – far more so than humans. In West Africa, the local mosquito species need the temperature to be within a range of 60°F–85°F (16°C–29°C). If the temperature goes below 50°F (10°C) or above 95°F (35°C), they die.

And they need those pools of standing water, too. So vegetation, soil, topography, and rainfall all have a part to play, as well as temperature – if there's no water, or if it's moving water, mosquitoes can't breed. And if the mosquito lays her eggs in a puddle, but it dries up before the eggs hatch, that's another failure.

So, in Southern Mali, where malaria is currently rampant, higher temperatures and lower rainfall resulting from climate change are both factors that will hinder mosquitoes. That means outbreaks are likely to be fewer and less serious. That's what researchers' modeling shows.

Though we should bear in mind that some big assumptions are being made in the climate model – for instance, that the world doesn't reduce its carbon footprint – and it's a very long-term model, running up to the end of the century. It looks likely Mali will still have to deal with hyperendemic malaria as a health issue for some time to come before the mosquitoes feel the pain.

Tipping the balance

Unfortunately, there are two sides to this coin. As West Africa benefits from drier, hotter conditions, East Africa will tip over into becoming far more attractive to the mosquito.

The Ethiopian Highlands, for instance, have always been protected from malaria by their elevation – and by the cool temperatures. Colder temperatures slow the incubation period of the malaria parasite inside the mosquito. In cooler areas, mosquitoes simply don't live long enough to become infectious. (Most mosquitoes only live a few weeks even in warmer areas.) But if the climate warms up, that won't be the case any longer (McSweeney, 2016).

The Highlands are some 1,200 meters above sea level. Though the landscape looks idyllic, this is one of the most heavily populated areas in Africa. If malaria is headed here, it's going to put a lot of lives at risk.

And now, the Highlands are heating up. MIT's climate model shows a 7°F (4°C) heat-up in the next few decades (Endo and Eltahir, 2020).

While this is by no means a wet country, the Highlands have always enjoyed more than enough rain to make a mosquito's life viable – over 800 mm annually. If the temperatures are high enough, the mosquitoes will not lack bodies of water to breed in.

The expected changes would put about a third of the region's population and 14 percent of its land area at very high risk of malaria within a century under a business-as-usual scenario (i.e., no cuts in carbon emissions).

The fact that the Highlands have so far been protected from malaria makes the diagnosis worse. In existing malarial areas, at least most people have built up some immunity to protect them. Highlanders have no such immunity, so malaria will hit them hard. Lack of acquired immunity not only increases any single individual's chance of getting malaria, but also increases the basic reproduction number – R0 – the number of infections a single person would be expected to generate.

At least we can use more sophisticated climate models than we had in the past. We can now predict the rainfall and temperature patterns that could lead to outbreaks of malaria, and focus preventative action where it's most needed. MIT's research shows that some areas of the Highlands are likely to remain low risk, whereas other areas – for instance, around Lake Tana and the town of Bahir Dar – will become high-risk areas for the first time. The landscape is by no means simple – a corrugated mass of mesas and mountain ridges where tiny farming compounds occupy hilltops, while cities like Mekelle, Gondar, and Arba Minch sprawl across the slopes.

* * * * *

Predictive analytics allow health authorities to focus effectively on very small areas. Previously, studies have mainly been done by Western subject-matter experts and handed over to local workers. But that means they can't be as precisely tied to specific sites and makes them difficult to update. Now Project DiSARM, with involvement from the University of California, San Francisco, gives African countries the predictive models they need, and the ability to update those models daily. Besides, increased smartphone use in Africa means it's easy to get locals to input good information – have they been bitten, do they use mosquito nets, when did it last rain (Noone, 2017)?

More precise models can help us to fight the disease with much more limited resources. Spraying a whole area with insecticide is expensive. Spraying a single village, or homes near a single site of standing water, could be all it takes to stop malaria from taking hold, and comes at a much lower cost.

Playing the long waiting game to see if climate change helps the fight against malaria in parts of West Africa is not ideal, especially when those very same changes may increase its prevalence in areas like the Ethiopian Highlands. Smart tech and advanced modeling give us the chance to play an active role in the battle to save people's lives against this deadly disease.

4

THE AMAZON RAINFOREST:
The Last Warriors

I T'S DARK AND HUMID HERE. The trees are thick, looped
together by vines; ferns and mosses grow dense and vivid
green. I hear sounds, sounds of animals I can't identify – a jaguar
coughing, perhaps? – of birds, the hum and buzz of insects. The
air is so dense under the tree canopy, I can hardly breathe.

Vines snaggle and tangle. My feet sink into the wet earth.

A snake humps its body over a log and heads off into the
undergrowth, its body spattered with a light and dark-brown pattern almost like army camouflage. I'm glad it didn't come closer.

For me, this is danger, the furthest I'm prepared to go; it's tor-
ture, clothes stuck to my body with my sweat. But for the Indigenous people, this is their homeland, their normal habitat. They
live off it and in it and look after it.

There's a shimmer of light far above; sometimes a ray of sun
penetrates through the branches for a moment. Nearer, something glitters – a hummingbird? A dragonfly?

I follow Taynara through the trees. Her face is decorated with
black lines and bright red; her hair is tied back by a woven head-

band. She calls herself a forest warrior. But she has her soft side; she stops and squats down to point at a trembling leaf, to show me a tiny, lurid-green tree frog with glowing orange eyes.

* * * * *

That's one side of the Caru river. Follow Taynara further through the jungle though, to the edge of the gray water, and you'll see something very different: a wasteland. On the other side of the river, the ground is bare, devastated by bulldozers, given over to grain or immense fields of monoculture – an agricultural model that the West is already giving up but which here is biting away chunks of rainforest every year.

It's green, but the pale green of puny earth-hugging plants, not the vivid diverse greens of the forest.

It's an ecological disaster. But for Taynara, it's her home that is being destroyed.

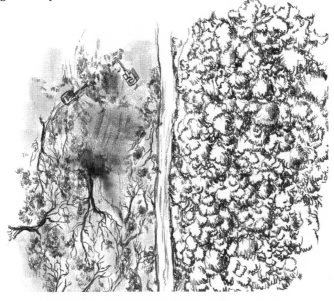

The lungs of the world

It's difficult to overstate the importance of the Amazon for the global ecosystem. It's the Earth's lungs – a huge carbon storage device, as well as a treasure house of still not completely explored biodiversity. And it controls rainfall on a large scale; it absorbs rain coming in from the Atlantic, which is sucked up by the trees; the roof canopy creates turbulence in the air, allowing it to absorb more moisture; and this moisture is circulated again and again as rain until it reaches the barrier of the Andes. If the forest starts to dry out, Amazon rainfall will reduce (welch, 2021).

To realize how interconnected the planet is, consider the fact that dust from the Sahara Desert is carried on the wind as far as the Amazon where it fertilizes the rainforest; and that moisture from the rainforest can affect rainfall as far north as the US. So, fighting for the Amazon isn't just about fighting for the rainforest – it's about fighting for the Earth as a whole.

* * * * *

To the 900,000 Indigenous people living in the Amazon, the rainforest is simply their home. It's where they get the herbs, tree bark, and other ingredients for their traditional medicine. It's where the fresh-water wells are, where they hunt for food, where their sacred places are. Each of the 305 different tribes has its own traditions, its own places; some of these tribes are isolated, while others trade along the river or with local towns, and even host tourists in homestays.

Most of them practice swidden (shifting) agriculture – burning a small patch of land that is enough to use for a few years, then moving on. Their herbal medicines have been examined by scientists, and some are now being used in Western pharmaceuticals;

they get material for their clothes, weapons, and housing from the forest – wood and vines, bark, and vegetable fibers. The Amazon is more than just a home, it's quite literally everything to them.

* * * * *

The Amazonian tribes are usually thought of as First Peoples living in a state of nature, but in fact, the Amazon is not an untouched landscape; archaeological research shows that people were living in communities here long before the Europeans arrived. They made pottery, and they cleared parts of the forest for agriculture. Over 10 percent of the Amazon is anthropogenic, showing signs of human management. But smallpox and other diseases brought by European invaders killed off much of the population, and some communities must have become unsustainable (Butler, 2005).

Until the late 20th century, there was no protection for the Amazonian tribes. But the Brazilian Constitution of 1988 established the protection of Indigenous territories, to the horror of the *ruralistas*, the agribusiness lobby which has been trying to overturn that protection ever since. Current Brazilian President, Jair Bolsonaro, is minded to help them do it.

For the *ruralistas*, the Amazon is a potential source of profit. For the Indigenous peoples, "The earth is our historian, our educator, the provider of food, medicine, clothing, and protection. She is the mother of our races." (Adventure-Life) The battle is not just a political one; it's for the future of the Indigenous peoples' cultures, and for the future of the planet.

* * * * *

The rainforest acts as the lungs of the Earth, a huge carbon-removal device that offsets the use of fossil-fuel energy. But deforestation threatens to shrink those lungs dramatically until the Earth can no longer breathe.

There are three stages of deforestation. First, there's the logging of the valuable old-growth tropical timbers. That takes the highest, oldest trees. Then the remaining wood is cut. And finally, the native vegetation is eliminated, as monoculture or grazing is introduced. (Badia i Dalmases, 2019) By this stage, the rainforest's original biodiversity has gone forever. Even if you plant trees again, you will never get back all of what was lost.

With swidden agriculture, the soil is used for just a few years, after which the forest is allowed to regrow for 10 to 50 years. People practicing this style of agriculture are semi-nomadic, using only small patches of forest, then moving on to another place. Regrowth takes much, much longer than the active use of the land (Adventure-Life).

But with modern monocultures, there is never a regrowth period. First come the logging trucks. Then come the cattle trucks. Then the soybean tankers.

And monoculture doesn't just destroy the forest. It kills the fish in the ponds, as fertilizers from the fields wash into the river.

The rainforest is also at risk from mining. Even though exploitation of Indigenous lands is barred by the constitution, there's little enforcement; illegal activities are widely ignored. And while Bill 191, an attempt to revoke Indigenous rights, was dismissed by Congress in June 2020, Bolsonaro is still trying to push it through.

IBAMA, the environmental police, collect only 10 percent of the fines they issue (Eaton, 2018) and are threatened by rioters, ambushes, and destruction of their vehicles. Their central government funding, already short, has been repeatedly cut – not just under Bolsonaro but under previous leaders too. And local people – often no more than youths trying to make a bit of pocket money – are often involved in relatively disorganized, low-level activity, like illegal marijuana farms or charcoal-making. The forest is under attack from all quarters, and its few protectors have little power.

It is quite astonishing to read some of Bolsonaro's comments about the Amazon's Indigenous peoples. In *Correio Braziliense*, 12 April 1998, he was quoted as saying, "It's a shame that the Brazilian cavalry hasn't been as efficient as the Americans, who exterminated the Indians." In *Campo Grande News*, 22 April 2015, he was cited as saying "The Indians do not speak our language, they do not have money, they do not have culture. They are native peoples. How did they manage to get 13 percent of the national territory?"

It's a populist theme that plays well with the *ruralistas* and their political machine – but Bolsonaro's program of clearing the Amazon rainforests has been referred to as "ethnocide". (In India, too, expansion of mining and logging often comes at the expense of Scheduled Tribes.)

You might think the age of colonization is over. But is it really?

System failure

Logging and mining are not the only threats. Warmer temperatures and increased carbon dioxide in the air are also changing the forest. The climate is becoming hotter and drier, conditions that are hostile to the Amazon vegetation that needs humidity to keep it healthy. And alarmingly, though the rate of deforestation had been headed downwards fairly consistently from 2004 to 2012, since then it's been on the rise again.

Under normal conditions, the Amazon works like a virtuous circle. More carbon dioxide makes the trees and other plants grow faster, so the forest can suck even more of it out of the atmosphere. However, the planet's main carbon sink has now become a net contributor to global warming (Eaton, 2018). How on earth did we manage that?

If you look at a map of land use, the dark green of pure forest has been nibbled away on the southern side and on the coast, by the pale yellow of agricultural land. In many areas, deforestation is already as high as 25 percent; some scientists believe if it goes past this point, it could be irreversible. Some even think half the forest could be lost by 2050 (Eaton, 2018).

When a forest is burned, the fire releases pollutants, including black carbon particles. These can trap heat and absorb sunlight, increasing the warmth of the atmosphere. (Welch, 2021) That affects the forest's environment and, if you like, its operating system; and the way the forest works is anything but simple. The rainforest circulates an enormous amount of water and energy with the atmosphere – and the water released by plants into the atmosphere through evapotranspiration, and to the ocean by the rivers, influences the global climate and the circulation of ocean currents.

But it's not a one-way process. It's a circular feedback mechanism – as global currents and weather patterns sustain the regional climate on which the rainforest depends (WWF).

Wetlands also have their place in the Amazon ecosystem, and as they dry out and are compacted through logging, they emit nitrous oxide – another greenhouse gas that can be 300 times more damaging to the atmosphere than carbon dioxide. If, on the other hand, they are overgrown with reeds and other vegetation, the plants help to reduce total nitrous oxide emissions.

Another greenhouse gas that is increasing in the Amazon basin is methane. Cattle ranching is a particularly big producer of methane, but so are flooding and dam building. While trees also release methane, their carbon dioxide absorption more than makes up for it in the global balance; but now, the balance is swinging the wrong way.

Even tourism, potentially a sustainable use of the Amazon due to no consumption of natural resources, brings plastic waste. Often, the families which find the plastic decide to burn it, but that just adds to the problem rather than solves it.

Compounding all the effects above, the Amazon rainforest is now a net contributor to global warming.

* * * * *

Could reforestation be a solution? It is certainly a part of it. For instance, agroforestry could be introduced, growing peach palms, teak, coffee, fruit, and vegetables in a relatively diverse polyculture. It's far more profitable per acre than ranching, and it also provides more jobs, so it's good for smaller farmers and farm laborers.

It could be very significant. Out of 79 countries with tropical rainforest, Brazil has more mitigation potential than 71 of the others combined (Loures, 2020).

It also reconciles the economic needs of local people with the needs of the forest, creating an economic future based on the idea of the forest as a productive asset.

But, unfortunately, reforestation is only half as effective as protecting intact forest (Griscom, 2020). Although reforestation may work in areas where logging has already destroyed the original ecosystem, the biggest single contribution Brazil can make in the fight against climate change is to protect the rainforest from further deforestation. Studies show that stopping deforestation can provide more than a third of the climate mitigation needed between now and 2030 to stabilize warming to below 3.6ºF (2ºC) (Rosamaria, 2020).

Rainforest warriors

Protecting the forest is the best way to restore the balance. And that's where the Indigenous people come in, because protecting their rights over their lands is the best way to protect the land itself. They're not looking for an NGO or a government agency to do it; they're doing it themselves. They are the Guardians of the Rainforest.

Historically it's been mainly men – sometimes with the involvement of women from their villages – who have armed themselves to protect their territories, or who have lobbied the government for their rights.

But the Guajajara women of Caru Indigenous Territory, in the state of Maranhão, have set up their own force, the rainforest warriors. Their territory, towards the northeast coast of Brazil, covers 428,000 acres (173,000 hectares) of primary forest. Why are these women "warriors"? Because they are mothers, they say. They are defending the land of their mothers, and they are defending it for their children. "If I am here today, I am the fruit of the women who came in front of me," said one, though she admitted that leaving her children at home to go on patrol is hard.

The force was established in 2014 and is still unique as a female-only force, though they work with male guardians on some of their operations. Some of the women had already accompanied their husbands on patrol, but creating the women-only unit gave them higher visibility and impact.

Indigenous women have the cards stacked against them. First, as women, they have already drawn the short straw. On average, even in developed countries, only one in four quotes in news media comes from a woman, and they're usually appealed to for "a personal view", not as subject-matter experts. Women often get left out of decisions on land use and environmental resources – look around a property developer's boardroom and you'll see how men still dominate the industry, even in the "progressive" West. The effect of this is multiplied by the fact that many charities and NGOs are happy to work with the (masculine) authorities already running things; less than 1 percent of philanthropy goes to women's environmental initiatives (Loures, 2020).

Second, they're indigenous – and Indigenous people, according to people like President Bolsonaro, are "savages". In fact, they're anything but. The women warriors run their campaigns using up-to-date technology. They take photographic records of what they find; they use satellite technology to find areas that are being deforested; they cooperate with law enforcement agencies; they have a drone and a quad bike. One of the young men in an Indian patrol turns out to be doing a degree in climate science (Badia i Dalmases, 2019); many of the women are highly educated on environmental issues. Maybe someone should tell Bolsonaro that many of them speak perfect, highly articulate Portuguese, and are eloquent in defense of their beloved forests.

(Incidentally, Bolsonaro is sexist as well as racist. He's well known in France for his dismissive comments about Brigitte Macron, wife of the French president.)

Given the forces stacked against them, these women deserve respect.

~ Based on photo by Marquinho Mota ~

* * * * *

Being a forest protector is dangerous. 140 environmental protectors have been killed since 2015; of 300 killings since 2009, only 14 went to trial (Loures, 2020). Sometimes, the only job the women have to do is deter local kids from growing pot – but when they're protesting illegal logging, they are up against large, well-funded businesses. They're also up against businesses that don't mind playing rough, and that are sometimes not too worried about legalities.

Even so, the fight for the forest is succeeding, at least in Caru. In 2016, the area lost 4,940 acres (2,000 hectares) to logging; by 2018 that had been reduced to just 156 acres (63 hectares).

Studies have shown that empowering Indigenous people with land rights and legal standing is the most effective way to help them protect their environment. From 2000 to 2012, deforestation in tenure-secure Indigenous lands was two to three times lower than in similar land without secure tenure in the rainforests of Bolivia, Brazil, and Colombia (Veit and Ding, 2016).

It's worth considering the stark fact that while Indigenous peoples and similar communities are responsible for managing up to 65 percent of the world's land, only 10 percent is recognized as legally belonging to them (Veit, 2016). (That increases to 18 percent if you include "designated" land). The best protectors of the forest are the people who have had it taken away from them.

And now, at least in Brazil, they're having to fight against the government as well.

* * * * *

The women warriors are making connections. They have united the 16 different Indigenous territories in Maranhão, where 200 women came together in 2017 to talk about protection, reforestation, and environmental education.

They aren't just warriors, they're also teachers, visiting communities on the river to explain their culture and way of life; and by doing so, they have gained a great deal of respect. They also provide lectures and training, within their communities, about how to fight for their own rights as well as those of the forest. And together with other groups, they're getting a very high profile in the fight against Bill 191.

* * * * *

If there's any doubt that Indigenous peoples are the best stewards of their land, take a look at Google Earth. If you overlay the Indigenous territories' borders on the satellite view, you can see that where Indigenous people live, there's less illegal logging and cultivation. The red borders of the Indigenous territories match, almost perfectly, the deepest green on the satellite view. Deforestation stops just where it meets the boundaries of Indigenous communities. That's a very simple, convincing way of seeing the difference. Global Forest Watch (GFW) puts it in numbers – while the state of Maranhão as a whole has lost almost 25 percent of its forest since 2002, the Caru Indigenous territory has lost only 4 percent (GFW, 2021).

In 2020, the Guajajara and other tribes decided to isolate to protect themselves from COVID-19. (COVID-19, by the way, was something Bolsonaro got badly wrong, calling it a hoax and no worse than the flu.) It's not a coincidence that, while they were not actively patrolling their territories, deforestation took off, reaching the highest level for 12 years (Welch, 2021).

Protecting the forest and protecting Indigenous peoples is one and the same thing. The Indigenous tribes are the best stewards of this precious resource, and any viable future for the Amazon rainforest needs to be planned around their stewardship and their economic needs. For instance, Fairtrade and some pioneer agriculture partnership projects employ Indigenous people to harvest Brazil nuts, and give them a much better price for their goods than they'd get from the usual middlemen.

The Amazon has such huge potential to help reduce global warming. Let's hope it can fulfill it.

* * * * *

A woman in a rainbow-feathered headdress. Her blue surgical mask doesn't hide the band of red make-up across her eyes.

Other women are wearing frocks and cardigans, but with multiple strands of black-and-white beads around their necks and feathered headbands keeping their hair tied back.

Three young men in body paint, huge beaded collars, and almost nothing else walk past. They bear banners reading FORA BOLSONARO (*Bolsonaro out*).

Behind them, the modernist concrete architecture of Brasilia makes a stark contrast. This capital was built for a future that hadn't allowed room for them – now they're taking it over, protesting the bill that could remove their rights.

But they're fighting for far more than just their own home – they're fighting for the entire planet.

PART I THINKING POINTS

A LTHOUGH THE GLOBAL IMPACT OF higher temperatures is frequently stated in the media, it's often difficult to grasp exactly what it means. "Two degrees hotter" doesn't sound like much. (That, of course, is an average – some areas will see bigger changes than others.) Or "sea level rising by 4 mm a year" – that doesn't seem a lot, particularly if you live somewhere where the tides rise and fall by several meters.

But it's undeniable that rising temperatures will threaten many people's way of life. Inuit will see much of their traditional fishing and hunting ranges disappear as sea ice melts and migration patterns change. French wine growers may have to change the centuries-old style of their wine as grapes mature earlier.

Listening to local people may be one way to help save the planet – and our cultures. In the Amazon and elsewhere, Indigenous peoples are often the best stewards of the environment, and perhaps it's time we listened to them, then gave them our full support.

Thinking points:

- What changes have you noticed in your life as a result of global warming? Are some foods less available or more expensive than they used to be? Have your air-conditioning bills increased? Have new species of insects invaded your house or garden, as tiger mosquitoes have in France?

- Are there low-impact ways to minimize these changes? For instance, some gardeners have decided to replace thirstier plants with Mediterranean varieties which need less water and prefer higher temperatures.

- You might want to think about how central a role consumption now plays in Western culture. Can we change that, for a more sustainable future? What values might be more sustainable?

- You might also think about how Fairtrade and other supply schemes can help sustain Indigenous, small-scale agriculture.

PART II

RISING WATER

5

MEKONG DELTA, VIETNAM:
Rice and Salt

THE MEKONG HAS SLOWED DOWN. As a young river, it carved great gorges and tight bends in the green hills of southern China. Now it's heavy with silt, slow with age, and has fanned out into different channels. It has only a hundred miles until it meets the sea, and it's weary, this great, wide, brown river.

Vietnam's Mekong Delta stretches over 15,000 square miles (nearly 40,000 square kilometers), made over six thousand years from the rich silt that the river leaves behind as it joins the ocean. It's a place of flat land and big skies; the river brown, and the paddy fields bright green as the rice grows, with irrigation channels gleaming between them. Long canoes with bright eyes painted on the prow ply the waterways; glossy black water buffalo pace sedately to the river's shores to graze and drink, and sometimes children come along for the ride. Storks and egrets hunt for frogs and eels in the mud, while cormorants dive for fish or sit on branches with their wings spread out to dry.

Many villages are mazes of tiny canals; going from one house to another can involve taking a "monkey bridge" made of a single tree trunk laid across the water. Locals run across without pausing for breath; tourists look long and hard before deciding whether to trust their sense of balance.

Old wooden boats cruise the river past small villages of palm-thatched houses; from time to time the spire of a Buddhist shrine can be seen, painted bright white or gentle blue, and the riverbanks are topped by coconut palms, spreading mango trees or the floppy, glossy leaves of banana plants. Or there are Cao Dai temples, vividly painted, like eclectic religious Disneylands.

Local hospitality comes in the form of sweet tea, or coffee made with a big spoonful of sugary condensed milk.

It seems as if nothing has ever changed here. Life goes on as it has for centuries, with just a few changes; some farmers have deserted their old wooden canoes for motorboats, and as elsewhere in Southeast Asia, motorbikes have taken over the roads.

In fact, life's not quite as natural as it looks. The delta has been shaped by engineering, infrastructure, and reclamation over the centuries – it's as man-made as the Dutch polders, so it's not surprising that Dutch experts have been working with the Vietnamese on the Mekong Delta Plan.

* * * * *

The Mekong Delta includes 13 provinces, inhabited by 17 million people. Even though Vietnam is now a fast-developing economy, agriculture still makes up a fifth of the country's gross domestic product and employs two-fifths of the Vietnamese workforce – even more if you include processing and agricultural supplies. And while other areas of Vietnam also have agriculture – such as coffee, most of which is grown in the central highlands – the delta is Vietnam's breadbasket, or rather, rice-cooker.

Rice accounts for over 80 percent of farmland in Vietnam, and the average Vietnamese gets 40 percent of their dietary protein from it. This is a country built on rice – and just over half its rice comes from the Mekong Delta (Troeh, 2015).

That's a bit of a headache for Vietnam. Because the Mekong Delta is terribly vulnerable. It's flat – there's hardly a hill more than 10 feet (3 meters) above sea level – so storms blowing in from offshore can drive seawater up into the delta and flood the low-lying land (Grootens, 2019). It's soft soil, not rock, so it erodes easily. Areas of the coast have been heavily eroded, some places losing up to 60 meters of land a year.

Some people have already seen their houses collapse as the banks give way, and in An Giang Province, the river has taken a big chunk out of National Highway 91.

If the sea rises a single meter (just over 3 feet), the delta is in trouble; seven million people out of the 17 million who live around the delta will have to move.

Agriculture is at risk too, even in the more stable areas inland. As the sea levels rise, salt water sweeps further and further up the delta, replacing the fresh river water and making rice cultivation impossible. The enfeebled Mekong has less power to push the sea-water out, so now the water up to 50 miles (80 kilometers) from the sea is salty. And for every 8-inch (20 cm) rise in sea level, salty water pushes about 19 miles (30 km) further inland (Vu, 2018).

Rice can only cope with 2 gpl (grams of salt in one liter of water) before it starts to suffer, and the water may be much saltier than that – up to 5 gpl, which will kill the rice completely.

Worse, the aquifers are depleted, taking away the main alternative source of water. The "Rice First" policy together with the growth of large towns in the delta have led to more and more groundwater extraction. The water table has sunk, and old wells no longer give fresh water.

The rice dies off first. But eventually, the salt affects other plants too; in many places now the coconut palms are dying off. The palm leaves turn brown, then they fall, leaving the trunks bare. Gradually, the trunks rot away and eventually tumble down.

There's also a massive human cost. Two hundred thousand households in the area are running short of water for themselves, before they even water the crops. No wonder some farmers are just giving up, with many poorer peasants migrating to larger towns or to Ho Chi Minh City, where they can find work.

The squeeze

It's not just global warming that menaces the Mekong Delta. It's in a squeeze. To the south, the sea level is rising; but to the north, up-river, China and other countries are building dams across the Mekong. Over the past decade, these dams have had a major impact.

First of all, the dams have reduced the amount of water that makes it all the way downstream to Laos. It's estimated that China's eight dams from Gongguoqiao to Mengsong have taken, in all, 40 billion cubic meters out of the normal flow that Vietnam would have expected before the dams were built. Laos, too, is building dams; in 2019 it opened the new 1.3-gigawatt Xayaburi Dam, and more are planned, despite warnings from the Mekong River Commission.

The impact was quickly seen downriver. Where the Mekong runs between Thailand and Laos, often forming the border between the two countries, Thai media reported sandbars becoming a problem for navigation, and irrigation channels running dry. Even further downriver, the effect is exacerbated by the current low levels of the Tonlé Sap lake in Cambodia, which feeds into the Mekong.

Not only is less water coming downstream, but the dams are also stopping the valuable nutrients that used to be washed down with the river water. Intensive farming takes the nutrients out of the soil, and because of the Vietnamese government's "Rice First" program, many farmers are now growing three crops a year instead of the traditional single crop – or at most two crops. Without the silt in the river water replenishing the lost, necessary nutrients, the soil is becoming impoverished and less fertile (Lovgren, 2019).

After the reunification of the country following the Vietnam War, the "Rice First" policy was introduced. Its aim: to get the country fed. In the 1970s, Vietnam was broke; it needed to become self-sufficient fast. But while Vietnam has developed over the last 50 years, the policy hasn't changed since. In the 1980s and 1990s, the focus on rice worked well – not only did the country become self-sufficient, but Vietnam also ended up challenging Thailand as the region's biggest rice exporter.

That gave the country the wealth to invest in technology and manufacturing. (Vietnam is now a favored destination for video-game developers' offshore software production.) But the "Rice First" policy still hasn't changed – Vietnam is a country with a commitment to long-term, top-down planning – and by the middle of the last decade, it was obvious that it wasn't working so well anymore.

There's another problem, too, besides the lower river levels and rising sea levels. Higher temperatures caused by the El Niño phenomenon have been heating the climate and reducing rainfall in Southeast Asia. In the 2014–16 El Niño event, drought and saltwater intrusion caused damage to agriculture and fisheries estimated at more than $3.6bn.

The earth dries out. Huge cracks open up in the fields, and the young rice plants start to shrivel and turn brown. In the past, if there was no rain, farmers could take water from the river. Now, drought gives them a dreadful choice: let the rice die slowly, or risk using salty water from the river.

Women are hit especially hard. In many parts of Vietnam, it's women who do much of the agricultural work – yet often, officials and aid agencies ignore them. Women-headed households are among the poorest – and can least afford to see their crops suffer.

Two ways forward

There are two ways to look at the future. One is the "big engineering fix" that treats the entire Mekong Delta as a single system. The other is a locally based, more differentiated response, listening to the farmers rather than imposing solutions from above.

The Vietnamese government's response over previous decades has been a commitment to major infrastructure spending. Great concrete sluice gates, huge cement embankments; dams and dikes are intended to keep the fresh water in and the salt water out. But because of rising sea levels, some of the older dams are already ineffective, and sluices can flush agrochemicals into the sea.

Worse, concrete seawalls can lead to worse erosion elsewhere as they direct the water forcefully towards unprotected, softer soil banks. But despite these shortcomings, the government has been going ahead with a huge water-engineering program.

On the other hand, local farmers understand their environment. A lot of them have switched from rice farming to shrimp farming. Shrimp can be immensely profitable; some farmers have been able to build themselves smart new mansions and villas on the proceeds of their new crop. Some earn as much as $100,000 a year, which is immense wealth in a country where the gross domestic product is still only $2,700 according to the World Bank, and average local salaries can be as low as $140 (Troeh, 2015).

And, as Bubba told Forrest Gump: "shrimp is the fruit of the sea. You can barbecue it, boil it, broil it, bake it, sauté it. Dey's uh, shrimp-kabobs, shrimp creole, shrimp gumbo. Pan-fried, deep-fried, stir-fried. There's pineapple shrimp, lemon shrimp, coconut shrimp, pepper shrimp, shrimp soup, shrimp stew, shrimp salad, shrimp and potatoes, shrimp burger, shrimp sandwich." There is a lot you can do with shrimp!

The farmers have managed to make their voices heard. Although Vietnam's economy is heavily controlled and centralized, farmers in the South were not collectivized as happened in North Vietnam. Vietnamese society is also highly consensual and rewards communal action. So when, in 2001, the government insisted on keeping the sluice gates closed, coastal communities which had switched their production from rice to shrimp protested. The women formed a human shield, while the men threatened to destroy the sluice if it wasn't opened.

In the end, rather than see the sluice destroyed, the government caved in. The shrimp farmers got what they wanted.

But the big plastic-lined ponds that contain the shrimp – while immensely profitable – are not an ideal solution to the Mekong Delta's problems, and perhaps not even to the farmers' problems. Many shrimp farmers see themselves living on borrowed time; instead of spending on developing their farms, they're trying to give their children the best education possible and telling them to move to the city for profitable work, rather than farming the family land.

Many of the shrimp farmers also end up heavily in debt. Starting up is expensive, and they have to make money fast to pay off their loans. That makes them wary of taking on any risk, and they worry about what next year might look like. For some, instead of being the solution, shrimp farming just becomes a new problem.

On the ecological level, shrimp farming is also a suboptimal solution. The ponds often replace forest or orchards; with no roots to bind the soil, and no windbreaks to protect the delta from increasingly common typhoons, the delta becomes ever more vulnerable. On the coast, shrimp farms have razed areas of mangrove swamp, which offer precious protection against tidal and storm erosion.

Twisty mangrove trees with their winding, snaking roots, make a much more subtle and more effective barrier against the sea than a concrete embankment. The roots trap earth and organic matter in between them and can help to raise the level of the soil by up to 25 cm a year.

The Dragon Institute, a research center for climate change, has been working with farmers to create a solution that works for both the farmers and the environment. The solution they've come up with for the coastal areas is not quite rewilding, but close.

Instead of building huge rectangular ponds for the shrimp, farming families are encouraged to build long, thin shrimp ponds, divided by mangrove swamp. There are benefits for everyone. The farmers, as long as they retain mangrove cover of 50 percent or more, can sell their shrimps as organic produce, getting a premium price. The overhanging mangrove also helps to keep the pools cool – overheating can kill the shrimp and wreck a farmer's profits.

Farmers further inland are taught about rainwater harvesting – a small-scale measure that can be taken by a single household or a whole village, but can help save a crop when the river water is too salty to use for irrigation.

There's also been a change in the way the delta is treated by the government. While the Vietnamese government traditionally looked at the whole Mekong Delta as a geographic unit, where rice should be the predominant crop, the Mekong Delta Plan

(MDP) looks at things differently. It sees the delta as not just a single environment, but an interconnected series of different ecospheres. Instead of trying to impose one solution, the MDP divides the area into different regions – the low-lying coastal mangrove swamp where sustainable aquaculture can develop within the existing environment, and the higher land further upstream that needs to be protected from saline intrusion and given access to new methods of water management.

The MDP also understands that farmers and local communities need to buy into whatever decisions are made.

So where do we go now? Despite the big hydraulic program, small-scale responses are already helping farmers in several provinces. The government is now allowing farmers in the central zone to abandon triple-rice-cropping and grow other crops – although fruit orchards are also suffering from saline water, they are not quite as sensitive as rice.

Meanwhile, the fresh-water floodplain that was taken over by intensive rice farming has been restored in the upper part of the delta. The idea is that if fresh water can be trapped in the upstream floodplain during the monsoon season, it forms a natural reservoir that can be called on when the dry season turns into a drought.

There are still a few problems to be sorted out. For instance, while the Mekong Delta is a single hydrological system, it spreads over 12 separate provinces. That's a dozen local governments as well as the central government in Hanoi that all need to buy into the scheme.

But the Netherlands Development Agency has extended its program, with the help of the International Union for the Conservation of Nature and the Netherlands Development Agency. Even locals in the Mississippi Delta have exchanged ideas with the Vietnamese and found a lot of common ground, including several great ways of cooking crab – and, you can bet your last dollar, shrimp!

6

FIJI:
Paradise at Risk

S TAND HERE ON THE BEACH and you can see how the water changes from bright turquoise in the shallows, to darker blues further out, broken only by a white line of foam where the waves break over the coral reef.

You feel the hugeness of the earth here; the curve of the horizon, the sweep of the beach, coves carved out of the island by the rise and fall of tide after tide; and the sunsets are wonderful – purple, gray, flushed pink, and a flash of hot gold as the sun sinks.

The sound of the sea is always in your ears, like a whispered commentary on daily life. And in this vast world, the island is just big enough for a couple of houses, a few coconut palms, and a little vegetable garden. Robinson Crusoe would have been happy here.

The nation of Fiji is made up of 333 islands, and many of them are just like this.

The center of the mainland is high mountain and cloud forest, but the people live along the coast. Once the sun goes down, there's singing, Polynesian feasts with meat cooking over hot coals, the spicy aroma of Indian curries and dals, or – for the tourists – cocktails in the beach bars. This is Paradise.

* * * * *

But Paradise is under threat. The sea is rising, and the tides are becoming more extreme. The people of Fiji have always known storms and cyclones, but these have become more devastating – Cyclone Winston in 2016 was a Category 5 (the highest on the scale), killing 44 people and affecting 40 percent of the Fijian population (Dougan, 2020).

It's difficult enough recovering from a single big storm. But an increasing number of smaller tidal surges bring salt water onto the land, destroying crops from excessive salinity, and contaminating drinking-water supplies.

In 2014, Vunidogoloa was the first village to be abandoned. Water was rising into the houses; the village beach had disappeared. Now, in the new village two kilometers inland, people are safe from the encroaching tide, but they feel empty – the old village is always in their minds (Dougan, 2020).

Vunidogoloa was only the first. 80 more villages will follow it.

It is the end of an era for Fiji.

* * * * *

There's a longer-term issue too – coral bleaching. Warmer seawater makes corals lose their color. And soon after they fade, they die.

A coral reef is an integrated ecosystem. When the corals die, the fishes and other marine animals who rely on them also die.

And fish are the main diet for the people – besides being one of the reasons that tourists come to Fiji for scuba diving and snorkeling.

In the traditional Fijian calendar, there's a fish for each month; but they can't find the right fish anymore. The fishermen say they always used to know where the fish would be; they could read the sea like a book. But the fish have become unpredictable, and rare.

And those lovely beaches that are one of the great attractions of Fiji are getting swallowed by the sea. In some villages, the first line of trees is beginning to fall. The local children want to play on the beach the way they used to – running to the first wave, then all the way back ahead of the incoming foam, the way children do the world over. But nowadays, all they can do is run along a thin strip of sand, there and back, there and back. It's not the same.

Cyclones: winds behaving badly

To understand Fiji's problems, you have to understand how cyclones form, usually in the summer and autumn. The sun heats the surface of the ocean, where the water can rise to temperatures of 82°F (28°C) or even higher (Met Office, 2021). This warms the air just above the surface of the ocean. As you know, warm air rises.

But that leaves a gap at sea level, which pulls in cooler air to replace the warm air that has risen. Then that warms and starts to rise, so the process keeps on moving, pulling in more and more cool air and pushing up more and more warm air.

As the warm air rises to higher altitude, it cools. That makes the moisture in it condense, forming water instead of vapor.

That's the normal process that creates precipitation, and in most cases, it results in no more than a short shower or a rainy afternoon. But in the case of a cyclone, it happens very fast and on a huge scale; the resulting clouds can be 10 km high. And when a cloud is that big, it feels the pull of gravity. The Earth's gravitational pull and its rotation (the Coriolis effect) makes it spiral – like water going down the drain after your bath.

You might call this a tropical cyclone, or a hurricane, or a typhoon. It's all the same thing with different names depending on where it's formed, a whirling wind demon that can move extremely fast, with huge destructive energy. Some cyclones expend all their energy at sea, so when they reach land, have died down; others hit the land with their full force.

Cyclones have happened throughout history. But now, with rising temperatures, conditions are becoming ever more favorable for cyclone formation. And when cyclones get going these days, they usually have more energy, too.

At the same time, rising sea levels have made Fiji and similar island nations more vulnerable to cyclone damage, since the cyclones can now push seawater much further inland.

And for the smaller islands of the Fijian archipelago, where most of the communities are coastal and infrastructure in the interior is not well developed, it's a particularly bad problem (Dougan, 2020).

* * * * *

The impact on Fiji's people is complex. Polynesian culture sees the ocean as a divine force with which the island people have a spiritual relationship, so a storm is not just physically devastating, it's also emotionally disturbing. One witness says: "I remember when the tsunami hit Samoa in 2013, the Elders said we were being chastised by the Mother of the oceans. If there is a hurricane that hits the Pacific, the Elders also refer as the Sisters are not happy with the people of the lands."

When people feel endangered, they lash out. In a patriarchal society like Fiji's that can lead to increased domestic violence, as Shamima Ali of the Fiji Women's Crisis Centre warns. According to Shamima, 64 percent of all Fijian women experience domestic violence, and despite a law passed in 2009, few cases are followed up or prosecuted; in fact, victims are often made to feel guilty for reporting an attack.

Ali says things got a lot worse after Cyclone Winston. Women living in evacuation centers – cramped conditions and shared accommodation – were sexually harassed. Some were raped; relatively few bothered to report the incident formally.

What makes it worse, she says, is that many women were actively managing to get dinner on the table for their families, and even rebuilt their houses single-handed, while men were still in shock and felt threatened by the women "taking over".

Fiji's society is very patriarchal, and women often have difficulty getting heard. That's why an important innovation has been creating networks that give women a voice. Eta Tuvuki, a fem-LINKPacific rural leader, interviews local women to see what challenges they are facing. One of the biggest issues after the cyclone? Finding clean water. Many women were having to walk miles to find a water source that hadn't been contaminated. Public village meetings, dominated by men, hadn't even mentioned the issue.

Eta Tuvuki says she is speaking up for "the shy moms, the shy ladies, the shy grandmothers, who cannot come out of their shell and share what they're going through." (Narang, 2018)

Salt-wrecked farmland is leading to increased poverty. But here, too, small steps are providing lasting solutions. As on other threatened coasts, growing mangroves has helped to provide a barrier against the waves, and polyculture farming such as agroforestry and intercropping has created a more sustainable form of agriculture. Some farmers now grow up to 30 different kinds of crops, diversifying so they don't get wiped out again: eggplant, sweet potato, mango, chilies, cassava, cowpeas, papayas, and more.

And contingency planning is a big thing in Fiji these days, from moms ensuring they have a ready-packed emergency bag of clothes and food to take with them if the family has to run from a cyclone, to farmers creating a market garden to supplement their main plantation (Narang, 2017). It's hard for a modern city dweller to imagine what it's like to be told you have five minutes to pack your bags, grab your kids or pets, and run for your life. Even harder to imagine that once the danger's gone, you might not have a house to go back to.

* * * * *

"What do we need? Blankets. Clothes for the children. Food. Grab some *dal* (lentils) and *roti* (bread).

The wind's roaring. It was low tide and suddenly the water's right at the door of the house. I can hardly hear myself think.

Rain's whipping through the doorway. How am I going to keep our food dry?

Come here, little one. Hold my hand. I can't have you wandering off to the beach. And Big Sister's at school. I hope they've moved up the hill already. I know they had a drill last year, but things don't always work as you expect . . .

No time to worry about that. Ah! The washing-up bowl. That'll do. And a bucket. Put the rice in there too. And I need some fresh water.

Oh, God. Look at the coconuts. They're bent over, they're going to break.

My feet are wet. The water's coming into the house.

Up on my back, kid. We've gotta get out. Up there, up the hill, up where the waves and the surf can't reach us."

* * * * *

One big step that's being taken by the Fijian government is to make climate change part of every Fijian's education. And it's working – it's young people who are leading the fight.

In school, Fijian children learn about the science behind the events confronting them. They experience first-hand the natural disasters that are happening – unlike many children in other places who only experience a warmer summer holiday. They see the flattened beach houses. They see palm trees stripped of their leaves, gaunt and black like strange, long-legged aliens. They see the land where nothing's going to grow this year, water-logged and salty.

And so Fijian youth has decided to take action.

Timoci Naulusala, at 11 years old, saw Cyclone Winston hit his village at 140 miles per hour (220 kilometers per hour). He saw water surging over the fields, wind bending the palms, metal sheets flying off roofs. The day after, he saw lumps of concrete in the mud where there were houses the day before.

In 2017 he made a speech at COP23, the UN Climate Change Conference, saying the world must take action – now. "Climate change is like a thief in the night," he warned.

AnnMary Vikatoria Raduva was a teenager when she joined the Global Climate Strike during the UN Climate Change Conference. "My generation and I are cleaning up the mess we did not create," she said (Dougan, 2020). She's gone on to speak at other conferences, recently joining one in Brisbane via Zoom.

~ Based on photo by AnnMary Raduva ~

But while high-level climate summits make the headlines, a lot of AnnMary's work and that of other young Fijians is being executed at the local level. AnnMary has campaigned against plastic pollution and balloon releasing; she's been involved in beach clean-ups and mangrove planting. (As well as stabilizing the soil, mangroves help absorb carbon dioxide, trap impurities, and stop land runoff washing onto the pristine coral reefs.) More than 6,000 trees were planted in just one year by local youth groups.

More recently, she's been putting together period dignity kits for women whose lives have been affected by COVID-19.

Some adults have been patronizing, but she says she's also received huge support – including that of her parents.

Adi Anasimeci Volitikoro is a keen soccer player. When she's not kicking a ball, she plants trees. She's only 19 but she's already leading a multi-racial, community outreach group and encouraging other youth groups to help the environment (Neimila, 2020).

Even the seniors of the environmental movement in Fiji are relatively young. Sivendra Michael was the first Fijian to graduate with a first-class economics degree from the University of Auckland. His doctoral studies assessed the resilience of small businesses in Fiji – from macro- to microeconomics. Now, with the Valuing Voices project, he's combining art and activism. And he's only just hit his 30s.

He describes himself as "passionate about sharing narratives" – and his motto is "small changes collectively make a significant impact." He believes that no one should feel helpless; that everyone has a part to play.

His "artivism" uses music, theatre, video, and social media to spread important messages. He also wants to open up artivism to everyone and ensure the full diversity of Fijian voices is recognized – people of both Polynesian and Indian descent, young and old, town and village dwellers, women as well as men.

Sivendra also helped lobby to increase the Environment and Climate Adaptation Levy (ECAL) from 6 percent to 10 percent and to invest it actively in environmental protection projects (Fullerton, 2019). But he warns against tokenistic use of young people; their voices need to be given a proper hearing, and one of the things that can help this happen is the development of networks for young environmental advocates. Young people led the campaign to ban single-use plastics in Fiji – and the government agreed to phase them out by 2020.

Hearing the voice of youth

What's clear is that when we are addressing climate change, the voices of women and the young need to be heard. We may need to change some of our institutions to make this happen, or to establish new networks outside those institutions. But bringing these different points of view to bear on the situation will enrich and deepen our response to climate change, and create real enthusiasm for the task of reducing carbon emissions.

Younger and smaller nations face the same issues of not being listened to – even though they're often the nations most affected by climate change. And like women and youth, they are rallying to make their voices heard.

The Pacific Climate Warriors held their inaugural summit in Fiji and made a strong statement: "We are not drowning, we are fighting". Smaller, less powerful nations have the same problems as young people – the big boys aren't always prepared to listen.

This organization will give them a stronger voice in the international arena.

With a 24-inch (60 cm) rise in sea level possible by the end of the century, it's not just lifestyle or wealth at stake; for some of the SIDS (Small Island Developing States), the question is whether they'll still exist at all.

That's one reason why the Pacific Islands pushed hard to get a supplementary 2.7°F (1.5°C) temperature goal in the Paris Agreement. It's an ambitious plan that aims to deliver a better result than the agreed, firm 3.6°F (2°C) goal.

As AnnMary says, these islanders are fighting for one simple reason: "Because my island, Pacific Island countries will be the first to go underwater and my people will become the first climate refugees." (Rovoi, 2019)

And that's why we should hear their voices.

7

BANGLADESH:
The Rivers Give and The Rivers Take

T HE WIND BELLIES THE YELLOW sail of a boat against
the deep-blue sky. The water ripples.

The net is full of silver. You can smell fish and salt; the ozone
smell of the sea.

Lush banana leaves graze the water. Thatched farmhouses
are shaded by tall trees, on islands where hens scratch in the dirt
and brown cows nuzzle with moist noses. Girls in bright orange
and green dresses chase along the top of the embankments, and
pick the bitter gourd, okra, and peppers that grow in their little
gardens. Fattening fish swish their tails in the dark waters of the
farm's pond. Elsewhere on the island, there are huge fields where
the young rice thrusts up green, shivering in the breeze.

Water was always wealth in Bangladesh. Water brought trade,
but more importantly, water brought the rich silt that made the
land so fertile; 525 million tons of silt a year. Sometimes you got
flooded, or a storm blew over and took your roof off, but that was
the price you paid for this fertile land where everything grew.

That was the price that Bangladesh paid for being the land of 700 rivers, a country two-thirds of whose surface is wetland, and where the average height above sea level is around 10 meters (Heitzman, 1989). This is where the Ganges and the Brahmaputra finally end their journey from the Himalayas; where, weary, they lay down their burden and gradually mix their waters with the salt water of the sea.

The rivers give, and the rivers take away.

Living on borrowed time

But Bangladesh is living on borrowed time. It is a densely populated country, with 165 million people living in almost 150,000 square kilometers – a bit larger than Greece, or, if you like, the size of Illinois or Iowa.

That means there are 1,260 people per square kilometer. That's a lot. The only countries with higher density populations are city islands or tiny enclaves – Singapore, Hong Kong, or Gibraltar, for instance. The USA is way down the bottom end with only 24 people per square kilometer. Bangladesh is a crowded country.

In the delta, many people live on islands made of compacted silt and sand, the *char* – a typical part of the southern Bangladeshi landscape. Most are just a few meters above sea level, and they are fragile environments. They used to last a few generations, but now the average life of a char is just over a decade (Thomson, 2017) as rising seawater and more frequent storms erode the chars' embankments.

Before, a storm passed over, then you got rebuilding. Patience, over the generations, paid off. But now, when a storm comes, some farms just disappear. There's nothing left to build on. In one case, a school was dismantled before the monsoon floods in the expectation that it could be rebuilt once the floods had gone down. Instead, the river undermined the banks and swept away the plot where the school once stood. It will have to be rebuilt elsewhere.

Imagine: this is the third time you've lost everything, and you have to rebuild your farm and your life all over again. And this time, when you go to the bank for what is, quite literally, seed capital to get a new crop started, the bank says no. Central risk management has done the sums; loans won't get repaid before the next disaster. For centuries, farmers here have been able to pick up the pieces and start again; now, they can't.

Cyclones kill thousands; one in 1970 killed up to half a million people. (No one knows the exact number.) But it's not the danger of a single great catastrophe that forces most people to move; it's the daily insecurity. A school child growing up on a char might remember three, four, even five house moves; too many for such a young life. For the parents, it's just too much work starting again every couple of years.

So, people are moving off the chars and into the cities. They may not know it, but they're climate migrants. There's a new pattern of migration into the towns, particularly the capital, Dhaka. In Bangladesh, 700,000 people a year are now on the move due to natural disasters, and a World Bank Report estimates the country could have 13.3 million internally displaced people by 2050. That would make nearly 10 percent of the entire population.

At the same time, Bangladesh has accepted more than a million Rohingya Muslims from Myanmar, who are fleeing discrimination and persecution. Living in improvised shelters around Cox's Bazar, they're adding to the total of displaced persons – it's a testament to the country's resilience that it's been able to cope.

But, until recently, there has been no government vision for how to deal with internal migration, other than ensuring that shantytown squatters don't get too comfortable – which means denying them power, clean water, or decent sewage facilities . . .

* * * * *

Everyone here looks out for everyone else. Little Fatima looks after her two sisters and baby brother, Ali, even though she's only a child herself; her mother works as a cleaner and cook.

If it weren't for the little stove, it would be dark in here, under the corrugated tin roof over damp concrete walls. It's not much – but it's home.

Next door are the Brickwork Boys. A dozen of them packed into a little hut, Fatima hears them get up before dawn. They come from all over Bangladesh, from villages she'd never heard of before, but they all work in the big brickworks. Sometimes, if they have a bit of dal and rice leftover, they offer it to Fatima's family.

It's a tough life in the brickyards but the boys prefer it to the alternative — working on construction sites. That could mean the 25th story of a skyscraper, and safety kit doesn't exist in Dhaka.

They helped Fatima prop her family's bed up on bricks when the monsoon came and the water started coming into the house. Now she ties baby Ali up in her clothes at night, so he doesn't fall off the bed and drown.

* * * * *

Let's back off a little and get some perspective. Let's look at Bangladesh from above and see just how this country fits into the interlocking geographies of South Asia.

The Brahmaputra and Ganges rivers are born in the Himalayas as rushing mountain rivers cutting their way through deep gorges. Fed by glacial melt, they slow down as they pass through India. The Brahmaputra becomes a great braided plait of a river in Assam, its many channels twisting their way across a great floodplain, and the Ganges becomes a fat, rolling river, though it still swells high when the monsoon comes.

Here, in lands shared between India's West Bengal and Bangladesh, these two rivers come together with the Tista and the Padma to form a huge delta. It's impossible, in the end, to know which river is which; they are broad and shallow, fewer rivers and more spaces between islands.

It's always been an unstable environment but now things are changing very fast. Far upstream, the glaciers are melting faster and faster, and disrupted rain patterns can swell the rivers too quickly. Meanwhile, the sea levels are rising, and in a flat country, it doesn't take much to tip the balance between seasonal inundation and being under water, period. Storms can reach further inland, washing away riverbanks and undermining the chars until the land slips away, leaving a cliff with a clump of silty soil at the base where there once was a field or a village.

One island, Bhola, has lost 10 kilometers off its southern tip in the last forty years (Bilak, 2019). It's still losing more. How long before the whole island disappears?

Rising sea levels bring rising salinity, too. Agriculture in the delta depends on the fresh water pushing its way down the twisting channels, keeping the salty sea at bay. Farming here involves striking a difficult balance. The monsoon rivers bring 140,000 cubic meters of water a second, but that shrinks to just 7,000 cubic meters a second in the dry season, so when should farmers work their land? When arable land is farmed in the rainy season, there is always the risk of flooding. And when it's worked in the dry season, there may not be enough water for the crops.

When you work to such tight tolerances and the equilibrium is so precarious, it doesn't take much to tip the balance.

And when half the total population of the country depends on agriculture – and that's 100 percent of the population on the chars – things can get very bad, very quickly, once that point is reached.

* * * * *

A lot of the land in Bangladesh is essentially *polder* – land at or even below sea level, surrounded by earthen embankments. So it's not surprising to find the Dutch development agency helping out. After all, the Dutch know all about polders. They created them. The very word is Dutch.

Coastal embankments have been set up, and machinery installed to pump out the water; sluices keep out the scouring tides, and more productive agriculture is the result. A key difference from the usual government or NGO scheme, though, is that this time, the decision was taken to pay local people rather than outside contractors to carry out the maintenance.

That worked. It gave people a sense of being involved in securing their future. It also led to social changes; for the first time, women were getting involved in work outside the home, which until then had been the preserve of men.

But it was only a five-year program. When the outside stimulus stopped, things became more difficult. The government was not maintaining major infrastructure such as sluice gates, fields were waterlogged again, and while older women had gained confidence through the scheme, once agricultural yields dropped, girls started being taken out of school because there was no money to pay for their education (Morrison, 2014).

Fortunately, a new program has now started – Blue Gold Bangladesh – with a total budget of $75.9 billion. And this program, instead of having one overarching idea, has backed a number of smaller, local schemes and enterprises. Some farmers are trialing pen fish, raising insect larvae as fish food, and developing low-cost tilapia feed. Others are creating small home-based ponds where fish can be fed on scraps from the family household – rice bran, kitchen waste, snails – and harvested in just a few months.

Farmers are adapting their crops, too. Chili peppers grow in a few months and can be harvested well before the next flood, whereas rice takes longer and is more vulnerable. Where fishponds have been set up, they're making the embankments into market gardens and growing new crops – bottle gourds, tomatoes, chilies, ridge gourds, sponge gourds, and leaf crops like spinach.

Rice hasn't been left behind, either. "Nation-building rice" has been developed by a Bangladeshi research institution using genetic modification, specifically to survive on the sandy soil of the chars.

A new, floating world

Some farms are now trying floating island beds. The women make tiny balls of cow dung and soil, each containing a vegetable seed, and look after the nursery until the seedlings are big enough to be transplanted. Meanwhile, the men create a sort of pillow or raft of water-hyacinth stalks and leaves, which floats just off the char. When the seedlings are ready, the men transplant them, and the rotting water hyacinth, together with the water, provides all the nutrients needed. There's no need for fertilizer.

Cucumbers and gourds of all kinds can be grown in this way. And at the end of its life, the floating farm is broken up and used as fertilizer for winter crops like turnips and cabbage. In one season, the farmer should be able to earn double the cost of making the island (Sunder, 2020). But it's not necessarily just about making money – one farmer says now that he knows he has enough to eat, he will share his crop freely with other families in his village.

In fact, these islands or floating farms are an ancient technique that's being rediscovered (Sunder, 2020), called *dhap* or *baira* – little rafts planted with vegetable seedlings. And what's particularly ingenious about the technique is that it uses a plant normally seen as an invasive weed.

It's a technique that only a few areas of Bangladesh used in the past, but now project managers believe it could be appropriate for up to 90 percent of Bangladeshi wetland farmers, adjusted to meet their circumstances. However, because the islands need relatively calm water, it may not suit every situation.

~ Based on photo by Amy Yee ~

Other trials include sandbar cropping, using compost-filled holes in the sand, and vertical gardening using stacks of compost bags, or giant bamboo and plastic sheet containers. Some aqua-geoponic systems that use floating cages to grow fish together with vegetables, and liminal schemes such as "dike gardens" that grow vegetation in sacks beside the river.

The floating world is coming to other areas of life too, not just agriculture. The Emirates Friendship Hospital is a ship that makes regular visits to outlying chars, and which provides health facilities including an eye clinic, dentistry, and gynecology. The NGO which runs the hospital also trains community medics – some of them women who never received much education, but are now learning, gaining confidence, and even thinking of taking further qualifications someday.

A further benefit to women from the program is that the female community medics can talk to young women about reproductive health, and provide sanitary protection that many girls on the chars find embarrassing to shop for.

Education comes by boat, too. For some char children, monsoon season used to mean that school was closed for the duration. Not anymore; now, there are 20 free school boats traveling the islands. Each boat is modeled on the traditional *noka* wooden boat; the first such school was started by a young Bangladeshi architect, Mohammed Rezwan, who went on to source funding from the Levi Strauss Foundation (LSF) and Bill & Melinda Gates Foundation for more boats.

Mohammed's foundation, Shidhulai Swanirvar Sangstha, now runs 20 free schools, providing education to as many as 70,000 children. It also runs libraries and adult education centers. The latter are focused on practical topics like increasing crop yields or making organic insecticide from neem tree leaves (Yee, 2013).

* * * * *

Bangladesh is truly a nation of rivers; the Bangladeshi Supreme Court has given all rivers legal rights. Even though it's a Muslim-majority nation, the idea of the river as mother is deeply engrained in its culture, as it is in India – which has given legal status to the Ganges and Yamuna, though not to its other rivers.

However, this may not necessarily be a good thing. It gives government agencies the right to evict people living by the rivers, and NGOs wishing to protect the rivers may not have the money to oppose corporations in order to protect the river's rights. But it's a start – as is the establishment of a National River Conservation Commission. Unfortunately, the Commission's budget is not big enough for enforcement of action against polluters and what Bangladeshi newspapers pithily call "river grabbers and encroachers".

* * * * *

Four thousand square miles of mangrove, the Sundarbans forest area is shared between India and Bangladesh. From above, you see a criss-cross pattern of squiggles as rivers meander, touch, and break into one another; it's as twisty as the mangrove roots which anchor the trees in the silt.

Mangroves are flood-tolerant, and their roots stabilize the soil, so that it's not easily washed away. Unlike a concrete embankment, which pushes the water away, the mangroves absorb tidal surges and storms, taking the edge off the water's advance.

Here, the waters teem with fish, and the trees are full of bee nests that provide succulent honey. The Bengal tiger has its home here too, which makes honey-gathering a dangerous occupation. Apparently, the trick of wearing a mask on the back of the head (tigers attack from the back, so a face ought to deter them) no longer works as well as it did. Maybe the tigers got smarter.

But illegal logging has removed some of the mangrove barrier, and increasing salinity is killing off other tree species which helped to create a viable forestry resource. Many farmers are giving up and heading for Dhaka.

Out of the frying pan, into the fire

Whatever happens to the Sundarbans and the chars, Bangladesh has a second problem to solve – the huge number of internally displaced people who have made their way to Dhaka.

In many cases, the shantytown community recreates that of the islands; people from the same char or the same district tend to make their homes together, re-establishing their village community within the slum. The slums are in some ways surprisingly orderly, with huts in neat rows, but they have big problems.

The government doesn't want to encourage shanty-town dwellers, so it refuses to provide infrastructure. There's no clean water, no sewage system, and no electricity (though some enterprising youths manage to hook up to the grid to steal energy). Fires happen regularly, as do floods, and the areas are unhealthily crowded; there are all kinds of biting insects, gastrointestinal diseases, and a very high infant-mortality rate.

If climate migrants fleeing the chars thought they'd get away from the floods here, they end up disappointed; the water can rise two meters above the floor of houses already built on stilts. Young children have been known to drown after falling off a bed or a walkway.

Yet this is also a problem of success. Dhaka prospered as the country turned from a 90-percent agricultural economy into one based on manufacturing and "offshoring" production of textiles and garments. There are skyscrapers at the end of slum streets; there are jobs here, unlike the chars, even if they don't pay much.

But the city's infrastructure hasn't kept up with its economic growth. There is no subway, no public-transport plan, no affordable-housing policy. And now Dhaka resists providing any service to the shantytowns, fearing that will make them permanent. Land tenure is uncertain. In other Bangladeshi towns, shanty-town dwellers face being relocated to make room for airports and other public projects. The migrants haven't escaped uncertainty, they've just replaced one worry with another.

This is not a small problem in Bangladesh. The residents of the skyscrapers are the minority; it is estimated that 60 percent of the entire urban population lives in slums (Jordan, 2017).

* * * * *

Fortunately, the government is waking up to what's happening. As well as trying to help char dwellers reclaim their lives, and defending the low-lying towns and villages, it has also realized that many villagers *will* lose their livelihoods on the land and need something better than a corrugated tin hut in Dhaka. There is now a policy of supporting secondary towns that can attract both displaced people and employers.

Mongla is one such town, a booming port town whose mayor has tried to make it a modern and sustainable place for new workers to live. Mongla has invested in flood control and clean water delivery – though only half the city is linked up so far, so there's more work to do. Pedestrian walkways with shade trees and even Muzak from hidden speakers try to get people walking rather than motorcycling, and factory and warehouse jobs are being created. The port, once faced with closure, is investing $710m to increase its container-handling capacity, and is now aiming to compete with the country's busiest port, Chittagong (McDonnell, 2019).

Smart planning which addresses climate issues and invests in jobs and public services could be the way forward – at least for Mongla.

* * * * *

Back on the chars, some villages are dying. Social cohesion, traditions, and families have split apart. A lot of women are left trying to run their farms and households without the men, who have traveled to the mainland looking for work; shops have closed down; some houses are falling apart after a few years' abandonment.

But though char dwellers have always been poor, they had a great sense of community. Can that same closeness be recovered elsewhere?

A joint UK-Bangladeshi research project, studying community-based adaptation to climate change, used the ancient Bengali storytelling tradition of the "Pot Gan" to tell the story of a shanty house and why its occupants had moved there. The Pot Gan played to diverse spectators, including the people of the shantytown themselves, and a climate-change conference. It won over every audience.

One scientist told how often he felt news about climate change was usually delivered to people in a top-down way, at a level of complexity they couldn't understand; it was a government message, a message from besuited men in offices. Instead, he said, the Pot Gan was about relating science to people at a level they can understand, and allowing them to express their opinions on the stories that are being told.

And that's powerful: "We as an audience became aware that we can change something" (Jordan, 2017).

Certainly, the audience in Dhaka's shantytown was engaged by the play. One of the characters arrived with a suitcase; he was broke, he owned nothing, he'd lost his house, he was looking for help.

A woman stepped forward from the audience. "I'm sure we can find him a home."

And the whole crowd joined in.

8

THE NETHERLANDS:
LIVING WITH THE WATER

F ROM THE OBSERVATION TOWER, YOU peer into the distance down a long, straight, road. Two carriageways heading in one direction, a grassy central reservation, and two carriageways in the opposite direction. On both sides of the strip is an expanse of gray water, below a pale-blue sky with wisps of white cloud.

The road stretches to the horizon, and you can't quite see where the water stops and the land begins again. On a sunny day, there's a cool beauty to this thin ribbon of road; on a stormy day, the closeness of the water is downright scary.

This is the Afsluitdijk. It runs for 32 kilometers (20 miles) and for 30 of those kilometers it's the only barrier between two huge bodies of water. It's such a thin line in the vast waters, you can hardly believe it will hold.

To the south, the water was once salty, but is now fresh; it was the Zuiderzee, or Southern Sea, but now it's called IJsselmeer – IJssel Lake. Since the Afsluitdijk was built, over half the IJsselmeer has been reclaimed as *polder* – rich farmland, much of it below sea level.

The Afsluitdijk is perhaps the most extreme symbol of the Dutch relationship with water.

* * * * *

The Dutch live in a low-lying country; the very name of the nation, "Nederland" (Netherlands) means "lowland". Here, three major European river systems – the Scheldt, the Meuse, and the Rhine – meet the sea. This has always been a delta country, partly composed of islands, later of polders. A quarter of the country is below sea level.

The sea has made Dutch history. It made the Dutch into fishermen. It made them colonial traders, with the Dutch East India Company (VOC) taking over Java and dominating the spice trade; they had links with China and Japan. And it made them experts at managing the tides and the water. They started reclaiming land from the sea as early as the 14th century and even went to England to help reclaim huge amounts of land in Norfolk and Lincolnshire. They built dams, dykes, sluices, and polders. If anyone excelled at managing water, it was the Dutch.

But storms and tidal surges always remained a danger. In 1953, for instance, a storm combined with a high tide killed nearly 2,000 people. The fight against the water is part of national identity – a pride expressed in the saying, "God created the world, but the Dutch created the Netherlands (Mostert, 2020)."

Things could have been a lot worse in 1953. Three million people's lives were at risk when the Groenendijk, one part of the seawall defenses along the river IJssel, started to fail. Attempts at reinforcing it were frenetic but unsuccessful. Then the mayor of Niuwerkerk had what seemed, at first sight, a completely idiotic idea – he commandeered a barge and ordered its owner to block the dike by driving the ship into the hole. It worked.

That's a testament to how the fight against the sea has formed the Dutch character. It has made the Dutch diligent, ingenious, and inventive, and it has influenced the cooperative nature of Dutch life – "all hands to the pump" when needed. It also informed a responsiveness to the environment, as a result of which, climate change – which you'd think would be terrifying to this sea-based nation – has created a huge new business sector in the Netherlands (Mostert, 2020).

Of particular interest is the recent shift in thinking by the Dutch regarding their key relationship with water. Instead of pushing water further and further away, they're creating "room for rivers".

Why?

When your levee system no longer holds water

Let's look at some of the science, or rather, the engineering.

Dams, dikes, levees, and polders are all connected parts of the Dutch system of water management. Dams hold water back. They raise the level on one side and create a reservoir.

Dikes and levees are similar in that they are intended to hold water back from the land. Dikes protect land that would otherwise be under the sea most of the time; levees are intended as protection from occasional flooding, to protect low-level land. For instance, you might build levees around a town, so when the river floods, the town is safe, and the rest of the floodplain takes the strain.

The definition of a polder is quite simple: low-lying land that has been reclaimed from the sea or a lake, using dikes to protect it.

Although such solutions have worked well ever since the Dutch started reclaiming land at the end of the Middle Ages, they're not quite as good as they look. They come with undesirable side effects. For instance, a system of levees is only as good as its weakest link; areas which have higher levees will be safe, but they will push flood water towards the lowest levee, and that's where the water will eventually – unless you're lucky – break through. (That's become a big problem along the Mississippi, where some towns are now engaged in a "war of the levees").

By channeling river water, levees also cause rivers to flow faster; and by constricting the river channel, they make the water rise higher than would be the case with a "natural" river with extensive marshes and shallows along its course. And they can create bottlenecks, where high levees that hold against floodwater further downstream create massive flooding upstream, where the banks are less protected.

The Dutch have realized that it's time to change. With rising water levels, if they keep building existing dikes and levees higher and higher, eventually they're going to be living behind 10-meter walls. The Room for the River program is being adopted as the country's new way forward. The centuries-old war with the water has ended. Peace has been declared.

* * * * *

Instead of building levees right next to the river, the Dutch are now pushing the levees back. Setback levees allow the river to create a floodplain that becomes a nature reserve. Reeds and marsh grasses grow to fill the space, waterfowl like the elegant great crested grebe sail happily on the waters, and even sea eagles and ospreys have been seen hunting the rivers.

Houses have been moved back from the banks. Just like the Californians building their houses closer and closer to the forest – then suffering from wildfires – in the Netherlands, until recently, towns edged as close to the river as possible. Not anymore.

Instead, cities are finding ways to create dual-purpose sites that give citizens space, but also provide spillover ways for the water. For instance, the Eendragtspolder rowing course is a popular water sports center, but if the Rotte river overflows (which it probably will do every decade or so) it provides a reservoir for the floodwater. City plazas and basketball courts are designed so that if there's a flood, they become huge water tanks.

In Nijmegen, where the Waal river makes a right-angled turn, 50 houses on the opposite bank of the river were bought up and demolished so that the levee could be moved a quarter of a mile further inland. That has allowed room for a second river channel which can store water when the river is high, as well as a wider floodplain that's given the town bike paths, walking and jogging routes, and new green spaces.

It works. When, in 2018, towns upriver on the Waal, in Germany, were flooded, Nijmegen stayed dry. The German towns have levees; Nijmegen has its floodplain. The Dutch experiment was vindicated.

Some people living on the lowest polders have also seen their houses bulldozed. The government has been buying up houses that are built in locations now considered perilous, and returning the land to nature. It takes time; people are paid the full market price for their property and given assistance to relocate. Just a few homes remain, high on mounds above the surrounding flooded fields. Marshland and floodplain can now soak up excess water, and the government can concentrate on protecting other areas of the country and strengthening its defenses (Baurick, 2020).

As usual, the Dutch are incorporating their environmental awareness into daily life. Children have free access to swimming pools if they can prove that they can keep afloat while wearing their clothes and shoes – just in case there's another flood like 1953. That's the same facet of Dutch character which puts cycle lanes and bicycle-parking facilities at the top of the list of a city's primary requirements. And individual householders are being encouraged to remove concrete pavement from their gardens and replace it with more permeable materials so the soil can help to absorb water.

* * * * *

Big engineering hasn't been completely abandoned. The Maeslantkering is an engineering marvel – a massive floodgate to protect the city of Rotterdam, which is the busiest seaport in Europe (and which, ironically for an environmentally minded city, depends on oil refineries for much of its business). Each arm of the Maeslantkering is longer than the Eiffel Tower would be if you laid it sideways. It's intelligent, too; a series of sensors measure the sea level and automatically close or open the barrier (Al, 2020).

And it has only ever been used once. (In 2007. It worked.) But for the last 20 years, it's been regularly tested – and when it's tested, watching the huge gates slowly close has become a big local attraction, often combined with a cycle ride and a picnic (Kimmelman, 2017).

Climate change experts

The dutch have also applied their diligence and ingenuity to solving other problems, including taking a leading role in attempting to reverse the processes of climate change.

You might think a small, densely populated delta country lacks the usual resources to become a big player in agriculture. Far from it. The Netherlands is the world's second largest exporter of agricultural produce, behind the US. Now, it's taking a leading role in sustainable farming – with a declared target of achieving twice the yield using half the resources. (If you make chemical pesticides, the Netherlands is a dreadful market to be in – hardly anyone uses them anymore.)

And though you'd probably think the Netherlands has more than enough water, Dutch farmers have also cut water dependency by 90 percent. Using greenhouses doesn't just keep plants warm, it also stops the evaporation of moisture from the soil – or from whatever other growing medium the farmer is using (often mineral wool). There are also plans to use heat from factories and geothermal sources to keep the greenhouses at the right temperature, further reducing the resources employed.

Some agricultural land will be lost to the Room for the River plans, but the Dutch are already adapting. A businessman in Rotterdam is developing plans for floating dairy farms. Other farmers use drones to get information on soil chemistry, nutrients, water content, and plant growth – precise measurements that let them adjust their actions to get a huge boost in yields (Viviano, 2017). Some are growing tomatoes in greenhouses on the top floor of former factories, and growing them 13 meters high, to take maximum advantage of the limited space.

Green-roofed buildings help insulate the interior while recycling carbon dioxide. Farmers buy pest-eating bugs and nematodes instead of insecticide – that's a huge business now – and even plan to feed livestock grasshoppers instead of grains. So your Edam or Gouda cheese could soon be grasshopper-fueled!

The Dutch are helping other countries with their climate change actions. Delta countries like Bangladesh and Vietnam are calling on Dutch expertise to help them work out the right way to adjust to global warming.

It's become a huge business. High-tech engineering, high-tech water management, adaptation to climate change; whatever they develop, the Dutch sell it to the rest of the world. Sometimes they give it away, too. At Wageningen University & Research, a world leader in food technology, many of the students come from developing countries and export not just ideas from the Netherlands, but the innovative, enterprising Dutch mindset.

Perhaps one of the things that's most unfathomable about the Dutch national character is that you cannot surprise or shock the Dutch. If you said, "Hey, let's feed the cows on grasshoppers!" in most countries, you'd get the response, "Are you crazy?" But the Dutch response is quite different: "Okay, that's an idea . . . let's think about it."

These are the kind of people who repair dikes by ramming them with a ship, after all.

PART II THINKING POINTS

R ISING WATER LEVELS ARE NOT only a major threat to those who live in low-lying areas, but also to agriculture. Higher sea levels bring high salinity, making some crops no longer sustainable. In some cases, this has already led to the internal displacement of the inhabitants of these areas – "climate migrants", or you could also say "climate refugees".

At the same time, ways of managing water that worked when sea levels were lower and the weather less extreme are now beginning to fail. Just building existing seawalls higher isn't the answer; in some cases, it makes things worse.

Listening to nature often makes life in the wetlands easier. For instance, in three of our case studies, encouraging the growth of mangrove swamps has helped to build a permeable barrier between the sea and the good land, stabilizing the coast while supporting biodiversity. Growing different crops, or simply growing them differently – with drone farmers in the Netherlands, for instance – can also help to reconcile agricultural life with the rising water level.

Thinking points:

- In Fiji, both women and young people have been at the forefront of environmental activism. Yet these are often the very people that NGOs and governments don't listen to. Is this a situation you recognize in your own life? How can we start to address it?

- What would a rise of a meter in sea level mean to you? If you live close to sea level, the impact is obvious. But even if you live higher up – where does your food come from? If it's produced in low-lying areas, prices could rise to reflect the increasing difficulty of growing it. Increasing salinity could mean clean drinking water becomes more difficult to find.

- Bangladesh already has a huge number of climate migrants whose villages and farms have disappeared under water. How is the world going to cope if whole nations are swallowed up by the ocean?

PART III

WILD WEATHER

9

INDIA:
THE INVISIBLE ENEMY

I HAVE AN APPOINTMENT AT a newspaper office in Calcutta. I dress carefully. Fresh suit, shirt. Neatly combed hair.

The heat and humidity hit me as soon as I leave the air-conditioned hotel. My shirt is clammy with sweat in two minutes. By the time I get to the newspaper office, I look like I've been through a washing machine backward.

But no one cares. There's a major fight going on between two of the journalists about – I can't believe this – who took the last cup of iced water.

"The monsoon is a week late," explains the editor. "Everyone's a bit scratchy. It takes the tiniest thing to set them off."

But this is the life of educated, privileged Indians. Elsewhere, a squabble over the water cooler is the least of it. In a New Delhi slum, a woman is waiting with her plastic cans and a big aluminum pot to get her water delivered. The running water's been shut off – everyone depends on the tanker for their water; for drinking, for cooking, and for bathing.

She can feel the heat of the road even through the soles of her sandals. The road surface is beginning to soften. Sometimes it bulges up in huge curves of hot black tar.

This is what a heat wave is like when you're not rich. And there's a fight breaking out here, too – but it's turning ugly, lots of pushing and shoving.

How hot is too hot?

India is urbanizing rapidly. Its population is growing fast – more than half the population is under the age of 25 – and more and more are headed to the big cities where jobs are easier to get. Some are MBAs working in air-conditioned offices and living in leafy colonies; others head for the slums and work as street vendors or rickshaw drivers. That puts pressure on the water supply, and in a heat wave, the city can become a very unfriendly place indeed.

And heat waves are becoming more frequent. And more intense. And longer, too.

Let's step back a moment and look at the definition of a heat wave. For the Indian government, there's an extremely precise definition: it's when the temperature is 8.1°F (4.5°C) higher than the average for the area for at least two days. If it's 11.5°F (6.4°C) above normal, it's a severe heat wave.

In 2010, there were 10 heat waves. By 2018, the number had increased to 484 – and 5,000 people died as a result (Chandra, 2019).

You think you know about heat? Try surviving in 124°F (51°C). That's how hot it got in Rajasthan, one of the northern Indian states, in June 2016.

Add on to that an expected rise of 7.9°F (4.4°C) by 2100, if we don't manage to reduce carbon emissions. "Business as usual" will see two to three times more heat waves, and each one could be twice as long as the heat waves of the 1970s through to 2005. That extreme heat of 2016 Rajasthan will be happening once every seven to ten years. Normally, you'd expect it to be a once-in-30-years event (Bibek, 2020).

* * * * *

Heat wave advice is the same everywhere – in France, in India, in the US, people are given the same message. Stay indoors, don't go out in the hottest part of the day, drink a lot of water.

The staying-in part? It's not that easy in India. In "traditional" India, most people work in agriculture. Remote working is not an option. And in "new" India, a huge workforce is employed in construction, building new houses and commercial properties in the ever-expanding cities.

In 2017, though heat-exposed work drove only a third of GDP growth, it employed nearly three-quarters of the entire workforce.

Compared to journalists, call-center workers, and other office and service staff working indoors, these outdoor workers typically have lower incomes. If they stay in, their pay gets docked – or they simply don't grow enough crops to feed themselves; but if they go out, they risk getting sunstroke. In the 2015 heat wave, taxi unions in Kolkata refused to work between 11 a.m. and 4 p.m. after two of their members died of heatstroke (Bibek, 2020).

India faces a decline in labor productivity if workers are too fatigued to do their jobs properly, or have to take a three- or four-hour break in the middle of the day. And that could add up to nearly 5 percent of GDP by 2030. That may not sound much, but in dollar terms, it's enormous, adding up to $150–250 billion.

The International Labor Organization (ILO), in its 2019 report *Working on a Warm Planet*, estimates that India could lose 34 million full-time jobs through heat stress by 2030. It's no surprise that construction and agriculture are the sectors that the ILO thinks will be the worst hit.

India also has a huge number of pavement dwellers – migrant laborers and families who have moved to the cities for work and simply camp out in a doorway, sleep in their cycle rickshaws, or set up basic plastic shelters or even cardboard huts under flyovers or on the edge of parks. Heat waves are hard on them; they have no "indoors" to go to.

Even making a journey can kill. There have been reports of frail and elderly passengers on Indian Railways dying in the overheated, overcrowded compartments.

* * * * *

Heat waves have also helped to create a specific problem for Indian agriculture: farmer suicides. It's a growing problem for India. Many farmers have got into debt through having to buy expensive fertilizers and new seeds while a succession of poor harvests reduced their income and their ability to pay the loan interest. Some actually drink fertilizer to poison themselves.

Until recently we only had anecdotal evidence and a few isolated figures. But a study from the University of California, Berkeley, dug deeper into the hard statistics. Researchers were able to trace a direct correlation between increasing temperatures during the growing season and increasing farmer suicides. (Higher temperatures *outside* the growing season didn't make any difference, so it's the impact on agriculture that is stressing the farmers, not just the heat.)

The change in temperature can be quite small. An increase of 1.8°F (1°C) on an average day during the growing season doesn't sound much. But it was linked to an increase of 67 in the number of suicides.

And the number of suicides as a whole is shocking. 300,000 Indian farmers have killed themselves since 1995 (Safi, 2018).

Unsurvivable heat waves

But while agriculture has its problems, the real issue for India is that heat waves are getting severe enough to threaten long-term human survival in some areas.

Let's take a look at the science of how human bodies are affected by a heat wave to understand how big a problem is stacking up here.

Normally, when we talk about climate change, we talk about regular temperatures, the kind you'll see on your thermometer on the wall outside. But when we talk about heat waves, we also need to look at what's called the "wet-bulb" temperature– the difference between the actual temperature a thermometer measures, and how hot or cold you *feel*. The wet-bulb temperature, unlike your thermometer, takes humidity into account.

In Calcutta, the local papers print the "Discomfort Index", a slightly different way of calculating the mix of temperature plus humidity; the index is the ratio between dry-bulb and wet-bulb temperatures. It's not just the high temperatures that put everyone in a grumpy mood – it's the stickiness, the sweatiness, the fact that even the notebooks journalists write on or the newspaper you're trying to read feel damp.

More important than discomfort, though, is the fact that the wet-bulb temperature reflects the ability of moisture to evaporate. We sweat – a big evolutionary advantage (dogs can't sweat, which is why they start to pant and drool when it gets hot). Sweat evaporates, thereby cooling down our bodies. (One old method of cooling beer barrels was to put wet towels on them – as long as the moisture levels were topped up regularly, the evaporation kept the beer cold.)

But at a wet-bulb temperature of 95°F (35°C), the human body can't cool itself down, since sweat can no longer evaporate due to the high air humidity. At 50 percent relative humidity, a heat wave of 113°F (45°C) is enough to kill – at 95 percent relative humidity (only a little higher than what I experienced in Calcutta), the temperature only needs to get as high as 96°F (35°C) to be deadly (Bibek, 2020).

If you add this information to the "business as usual" projected rise in heat wave temperatures and frequency, parts of South Asia will become too hot for people to live. The McKinsey Global Institute has already warned that large cities in India could be among the first places in the world to experience heat waves at temperatures above the survivability threshold (Woetzel *et al.*, 2020). While rural areas will still experience heat waves, it will be in the energy-intensive, concreted-over cities that temperatures will be highest.

You may think global warming of a couple of degrees can't do this. But a small change in the overall average can lead to a massive change at the extremes, if you consider how temperatures are distributed statistically across the world.

Temperatures form a standard distribution – a bell curve, like the side of a tricorn hat. By far the majority of temperatures are clustered around the middle, with just a few outliers to the edges – the very cold and the very hot. Suppose you draw a line at 96°F (36°C). Right now, it's at the very thin end of the hat. If the whole curve moves slightly to the right – if you move the hat – more of the curve has now crossed the 96°F (36°C) line. You only need a slight move in the middle of the curve to make heat waves far more common.

Princeton researchers say that every 1°C (1.8°F) increase in average temperatures could increase the number of heat wave days by 4 to 34 days every summer. If the Earth warms by 5°C (9°F), as it could by 2100, tropical regions could get 120 days of heat wave – a third of the entire year (Berardelli, 2019).

* * * * *

But we've got air-conditioning, right?

Not so fast. At the moment, only about 10 percent of Indians have air-conditioning. It may well be that by 2050, most people in India will have an air-conditioning unit – but unless some hard work goes into reducing the units' carbon footprint, that's not going to be the optimal solution. In fact, it will probably make things worse.

Modern cities can create what's known as an urban heat island (UHI). Concrete, tarmac roads, and other building materials act as storage heaters. A major city, with 1 million people or more, can have temperatures 1–3°C warmer than surrounding rural or suburban areas. The effect is particularly bad at night – when the difference can be up to 12°C (22°F) – which means that city-dwellers who have been suffering in a heat wave all day don't even get relief from cooler temperatures at night.

Living in a UHI is particularly onerous for those in low-income areas, where there's limited mechanical ventilation, the population is dense, and there are relatively few trees.

And when you use your air-conditioning – which the growing Indian middle class increasingly does – you are not cooling anything down, you are just moving the heat around. So your own house, office, or hotel room is nicely cool – but just outside, your air conditioner is emitting even more heat into an already baking atmosphere.

You may not think this is a big problem. But India's booming generation of graduate workers in IT, engineering, and medicine all want air-con. Their parents did without – they're not prepared to. In one electrical appliance store, 30 percent of shoppers were first-time air-conditioning customers (Frayer, 2019). That rapid increase leads to predictions that by 2022, India will own 25 percent of the world's air-conditioning units. And they will all be pushing out extra heat into the streets of the big cities.

No wonder an LSE Grantham Institute study in 2019 warned that Delhi, Mumbai, and Kolkata were already getting close to dangerous heat wave levels. Of India's big cities, only Bengaluru was still in the "warm" rather than "hot" or "very hot" category.

India is rapidly urbanizing, following the same trajectory that Western Europe did in the 19th century. By 2018, 34 percent of the population lived in cities. India already has five megacities (population of over 10 million) – Delhi, Kolkata, Chennai, Bengaluru, and Mumbai. By 2030, Ahmedabad and Hyderabad will probably have joined them.

Traditional cooling, simple measures

India used to have its own natural air-conditioning systems. For instance, Bengaluru City was designed around hundreds of artificial lakes and pools, made by damming streams with bunds, or dikes. The lakes had dozens of uses – they provided drinking water, irrigation, fish, and somewhere to wash clothes, and they also helped to cool the city's atmosphere. But the number of lakes and tanks fell from nearly a thousand in the 1960s to fewer than 580 in the 1990s, and many of the larger lakes have been encroached on, due to increasing urbanization and skyrocketing property prices.

In the state of Gujarat, and across much of northern India, stepwells were built – huge staircases dug through the bedrock to access groundwater. It seems a rather extreme way of accessing water supplies, compared to simply digging a well and using a bucket or pump to bring up the water – but that's not the point. These architecturally impressive and often highly decorated stepwells contained spaces where women in particular could spend time during the hottest part of the day in cool surroundings.

That's how India used to manage heat. What's the answer now?

Some very simple measures can help to reduce heat wave deaths. The problem with heat waves is that the damage is invisible. Yes, it's hot, but you can't see the danger in the way that you can see a hurricane, or a flood, or a landslide. Many people don't notice the difference between hot, very hot, and unsurvivably hot, until it's already too late.

So, communicating the news to people is incredibly important, together with advising them on simple safety guidelines – how much water to drink, for instance, and how to keep cool. In some areas of India, local governments shorten school days and public sector working hours as soon as a heat wave is declared. In many cities, city gardens and parks are opened up, and people are told to go there for the shade.

There has also been a training campaign to help hospital staff identify and treat heatstroke. For instance, dizziness can be a symptom – so can severe diarrhea. Patients with heatstroke no longer have to wait in stuffy corridors or entrance halls for triage, but go directly to cool rooms, where fans, ice packs, and air-conditioning help them recover (Safi 2018).

In Ahmedabad, the capital of the state of Gujarat, a state action plan was introduced in 2013. Extra drinking stations are created during heat waves, and text messages are sent out to notify people they need to stay cool – and how to do so (Chandra, 2019). WhatsApp – one of the most popular social media platforms in India – is also used, as well as news outlets, and temples, mosques, libraries, and other public buildings are opened up as cooling centers. Outdoor workers are handed ice packs to help keep them cool (Ramirez, 2020).

This example has been followed by other cities and even some states such as Odisha and Maharashtra with Heat Action Plans.

In some neighborhoods, distributing free drinking water helps. Not every area in India has clean drinking water; supplies are jealously guarded. In some areas, a single standpipe supplies an entire neighborhood – and it's stopped working; in other places, criminal entrepreneurs have found a way to puncture the drinking water pipelines and steal the water to sell at a profit. A free source of clean water encourages people to drink enough to keep their body heat out of the danger zone.

While India's air-conditioning addiction continues, slum dwellers have found an effective low-energy, low-tech way to reduce the temperature inside their homes; simply painting the roof white. By reflecting the sun's rays, white paint can reduce the temperature inside by up to 5°F (2.8°C).

In 2014, an estimated 800 people's lives were saved through these low-cost measures. While not a huge number for a population of 1.4 billion, that's 800 individuals with families, friends, employers, and customers who need them – people who would miss them if they were no longer here.

The great thing about these low-tech measures is that they are so simple to apply at a grassroots level. Simply creating safe, comfortable green spaces within neighborhoods can both help individuals keep cool, and pull down city temperatures, reducing the urban heat island effect. Creating avenues for wind flow within cities, planting avenues with trees and other plants, and creating fountains and watercourses, are all small ways of greening the city and keeping the heat down. Ahmedabad is already leading the way – it expects to plant half a million trees a year over the next five years (Ramirez, 2020). That can make a huge difference, when you think that simply sheltering from the sun under a tree can make the air feel up to 27°F (15°C) cooler on the skin.

India is still a big fossil fuel consumer, but recent moves to invest in solar power could reduce its carbon footprint despite high economic and population growth. Meanwhile, let's hope the rest of us can pull together to reduce our carbon footprint as well and stop things from getting worse.

At least one guy in Kolkata is not too unhappy about the latest heat wave. A typical enterprising Indian, he built his own tea cart out of pallet wood and old car parts; he now finds he does a fast trade in all kinds of cold drinks too.

10

AFGHANISTAN:
BETWEEN THE TWO FIGHTS

T HE VALLEY SINGS WITH LUSH green fields and a braided turquoise river between the bare rocks of the mountains. Even the rock is tinted with red, purple, orange, brown. Further upriver, willow trees shade the fields with their tiny bright leaves, and tall poplars march, darker, along the riverbank.

The sky is pastel blue, the furthest hills are azure with distance, and the mountaintops are dazzling with snow.

In spring the valleys are bright with flowers: tulips, violets, and irises; red, purple, yellow.

The vibrant red mountains of Helmand Province flicker in the sunlight as if on fire.

In the villages, little girls in colorful clothes, their eyes ringed with dark kohl, still have that steady, serious look of young children. But their mothers, in flowery, floaty *shalwar kameez* (long tunic and trousers), or vividly embroidered dark robes with bright silver jewelry, break into shy smiles, or laugh as their children and grandchildren play outside the house.

In the fields, a gray-bearded farmer wearing a dark turban digs the ground with a mattock. His apple trees are already white with flowers.

This is Afghanistan.

* * * * *

Most of us have a picture of Afghanistan that is considerably less beautiful; a war-torn desert.

For many Afghans, too, that is their life; internally displaced refugees, living in tents or shacks hastily built out of mud bricks and scrap materials. Forty years of conflict have wrecked the country's infrastructure; the country hasn't known real peace since 1978 (Jones, 2020).

First, there was a war against the USSR-backed Marxist-Leninist government of the People's Democratic Party of Afghanistan, then a full-scale civil war, then the rise of the Taliban, and finally US-backed regime change – despite which, the Taliban and other forces have continued to fight, particularly in the provinces.

Under the Taliban, strict Islamic law was imposed. Women were made to wear the burqa, girls were forbidden to go to school, music was banned. (In neighboring Pakistan, Taliban hit men shot and wounded Malala Yousafzai, a teenage activist for girls' education; she, like many Afghans, is a Pashtun. Having been granted asylum in the UK, she continues to fight for the education of girls and women.)

Car bombs, massacres, assassinations – that's life in much of Afghanistan. Even where there is peace, there's still poverty; there is little fertile land, even less than there was 40 years ago. Deforestation, drought, and floods have wrought havoc on remote communities.

That's what links these two sides of Afghanistan: the ecology of the high Hindu Kush mountain range, and the effects on it of climate change and poor landscape management. When climate change makes water a scarce commodity and dries up the crops farmers have been growing for generations, conflict begins – and it's easy for organizations like the Taliban to exploit it. Then, while that conflict is going on, it becomes next to impossible to do anything to ameliorate the situation.

Dried out, then flooded out

Droughts and floods are both problems affecting Afghanistan. Again, although they may look like two distinct problems, they're two sides of the same coin, increasing imbalances within, and disruption of, the ecosystem.

More water evaporates from the soil during times of drought. Dry soil becomes vulnerable; it can just blow away in strong winds. If this is combined with low plant cover, particularly deforestation, and land degradation owing to overuse or conflict, you're trapped in a vicious spiral.

Drought comes with higher temperatures, and warmer air holds greater humidity. Rain only falls when the air cools enough for the water vapor to condense into liquid form. So, with higher temperatures, there is less rain for a long period, causing a drought – but because of the increased amount of water vapor in the air, when the rain finally does come, it can be overwhelming.

And for every 1.8°F (1°C) increase in temperature, the atmosphere can hold 7 percent more moisture (Carey, 2011). Add that together with other inputs to the climate model, and while moderate rainfall will only increase by 2–3 percent, extreme rainfall will increase by 6–7 percent.

People try to convert forest areas to farmland or use the wood for heating in the severe winters. The latter route leaves more land unfarmed. In other areas, conflict means farmers have left the land uncultivated. This feeds back into the system by warming the air further, which suppresses rainfall even more.

When rainfall finally comes, the eroded soil can't cope. Deforestation leads to higher chances of landslides. (Forests and scrub are a natural protection against floods; remove them, and you take away the protection.)

High-altitude farms often depend on glacial meltwater for their needs – a slow, regular drip resulting in small rushing streams that can be controlled by small dams and water channels. But the International Centre for Integrated Mountain Development (ICIMOD) says that glacier volumes in the Hindu Kush, the main chain of mountains linking Afghanistan and Pakistan, will decline by 90 percent by 2100. Thawing will be faster than the steady drip feed farmers are used to, creating uncontrollable surges of water.

With floods, droughts, landslides, earthquakes, and avalanches, Afghanistan is now subject to extremes of weather, and its inhabitants seem doomed to misery. It's a complex, diverse landscape, but in common with other areas of southern Asia, it has been increasingly subject to deforestation; in the 19th century, 50 percent of the land was covered by woodland – now it's just 2 percent. In 1978, 68 percent of the land was used for arable farming; now, only 16 percent is cultivated. The land is drying up; 1998–2006 saw eight successive years of drought, and out of the subsequent 11 years, eight saw both droughts and floods. Drought is expected to be the norm by 2030, and much of the country could end up as desert (UNDP).

Around Herat, Afghanistan's third largest city, the earth is fine beige dust; shantytowns of tents and mud-brick shelters house refugees from other parts of the country. In the shantytown around the capital, Kabul, it's just as dry, and there's no access to clean water anywhere.

Some farmers can't afford to keep going; others get into debt they can't repay. In some cases, the friendly businessman who lent money last year turns up again demanding to be repaid with a wife – for himself or for his baby son. Farmers end up selling their daughters, who are often too young to understand what's happening to them. Government cash handouts, intended to get farmers going again, end up being spent on paying off debt, instead of stopping the vicious cycle.

Due to its incredibly diverse landscape, Afghanistan can experience drought in one place, and flood in another – or drought followed by flood, a killer combination. When the glaciers melt too fast, there are avalanches in the high valleys; natural dams can be broken, then flash floods tear downhill, scouring the soil away, until a huge brown wave rushes through villages destroying land, crops, and homes. Topsoil and terraces are swept away (Jones, 2020).

In 2018 a glacial lake burst its banks. The flood rushed into the valley of the Panjshir River, setting off landslides that destroyed towns and fields. The normally vivid turquoise Panjshir was dark with silt; the ice melt that usually brings water for the fields, the families, and their animals, brought death instead. Villagers saw their houses crumble into the flood and their fields waterlogged; they left for good – many of them for Herat. One man woke in the night to hear the rumble of the water and managed to save himself, but all 10 members of his family were killed. Later, villagers had to search for the bodies; smashed against rocks, half-buried by mud, or thrown into the branches of trees (Mehrdad, 2020).

~ Based on photo by Silvia Alessi ~

Environmental degradation and conflict are not separate problems; one exacerbates the other. The UN Environment Programme says that 80 percent of conflicts are about land, water, and resources. That's particularly the case in Afghanistan, where agriculture is practically the only way to make a living, outside the main cities. Every family is a farming family; every family depends on the land.

Nearly a third of Afghans are now either internally displaced, in the camps surrounding the cities, or have emigrated. Men often go to work in Iran, leaving their families behind. It's difficult for women who are left having to keep their families together in a patriarchal society. One woman who was rich by the standards of her village lost her three cows and all 50 of her sheep in a flood; she ended up in a camp.

Girls are particularly affected by agricultural disasters. When there was drought in Bamiyan, a fifth of all girls in the local schools were pulled out of classes; they had to go to collect water or work in the fields. In other places, parents didn't have the money to let children continue their studies. Other girls were married off to settle debts. One girl in such a situation was so miserable with her abusive husband, she tried to burn herself to death a year after the marriage.

* * * * *

Women's rights are not an issue the Taliban cares about – and the Taliban, by some estimates, controls 70 percent of the country. On the other hand, the Taliban, like any military force, needs funding to maintain control, so it has become a major player in the drug trade. Farmers in some regions are growing opium poppies; the plants are drought resistant and much more profitable than anything else they could grow. The drugs trade in Afghanistan is now estimated to be worth $60 billion – as much as the country's entire official GDP.

~ Based on photo by Mai Ai Bing ~

Of course, the US Army has tried to stop the opium trade, often by dropping defoliant and pesticides on the poppy fields. That not only kills the poppies but ruins the land for anything else and puts harmful chemicals into the environment.

Illegal drugs also cause conflict between armed gangs or between US and Taliban forces. Women in the camps tell how their children used to snuggle up for comfort when they heard gunfire in the night; they've come to Herat or Kabul because they daren't go home.

For the men, food insecurity is a good reason to join the Taliban forces or work for one of the warlords who run particular regions. If you don't know where your next meal is coming from, joining up is a way to ensure you get fed.

(The definition of extreme food insecurity, by the way, is surviving on *less* than one meal a day. It applies to 13.5 million people in Afghanistan, almost doubled since 2017.)

And instead of fighting only in the summer, less snow and higher temperatures make it easier for the Taliban to fight all winter. Again, the environment exacerbates conflict.

* * * * *

Is there a way out?

The government talks big, but has done relatively little. The Ministry of Energy and Water has been talking about big dams, but these create major security risks, and, potentially, diplomatic issues with neighboring countries, so most development agencies no longer favor large-scale dam projects.

And too often the US, despite its "hearts and minds" mantra, went for quick-fix projects that weren't going to help anyone in the long term. Smaller dams have often been made by villages but aren't strong enough to withstand flash floods and landslides.

Meanwhile, US money has gone to the army. The US military operation in Afghanistan has had $745 billion in funding since 2001, but very little has gone towards infrastructure or agriculture (Jones, 2020).

So Big Development isn't working.

Small seeds of hope

But at a lower level, things *are* changing. It's a pity the UN is only able to put $71 million into new approaches to agriculture – a pittance, compared to what the US spent on armed conflict – but its small scale, local-level projects are transforming farmers' lives – particularly those of women.

At Pilawary Farm, Kabul, 100 women have been employed on a 16-hectare organic farm. They are not just growing food, but also processing, packaging, and marketing it. Processing adds value and helps create more jobs – and it can improve the quality of the product. Near Kabul, solar dryers are being used instead of simply spreading out food to dry on the ground, where it gets dusty and dirty. There are similar projects in other areas of Afghanistan.

A lot of thought has gone into finding other drought-resistant crops so that farmers can move away from opium. Saffron crocus flowers, whose pistils are dried and used as a spice, are highly profitable. (Saffron is the world's most expensive spice, at up to $72 per ounce.) Using greenhouses helps to increase the productivity of farms and retain moisture in the soil. What's more, growing gourds and tomatoes is more profitable than arable farming.

Even just encouraging farmers to plant fruit orchards and cultivate their vegetables in the shade of trees can make a huge difference, by stabilizing the ground and providing cover. It's a second source of income, and meanwhile, farmers can intercrop potatoes, pumpkins, carrots, onions, or eggplants. Compared to traditional crops of wheat or barley, it's more profitable.

Beekeeping is also an option that many women have taken up. It requires less labor and creates a higher income than arable farming, but farmers need help to understand how to keep their colonies productive and in good health. For some, it brings a whole new meaning to life. A widow's life in Afghanistan can be miserable, but one widow who took up beekeeping says the bees are her best friends – they make her happy, as well as making her income.

Even new heating systems, such as biogas plants that are fueled by cowpats, can help both the family and the environment by providing clean gas as well as organic fertilizer. Farmers no longer have to chop down trees to get through the winter.

As well as this, there are small-scale agriculture schemes that make sense – better water management, for instance, by retaining spring floodwaters behind dams, or water harvesting, together with early warning systems for floods. In some villages, the ancient systems of underground channels, known as *qanat* or *karez*, bring water from mountain springs without evaporation. These channels are an ancient part of Persian, Arab, and Afghan tradition, showing, as is often the case, that sometimes the old ways are the best.

At the same time, smaller canals are being created to distribute water more evenly, and contour bunds are being built, which slow down water runoff and help to control soil erosion. Community dams are now constructed with the aim of ensuring floodwater bypasses villages, rather than holding large volumes of water back. Often, a series of small check dams rather than one large barrage proves the most efficient system.

Farmers are also being encouraged to plant more trees – though this could take decades to become really effective.

One big change – and one that Malala Yousafzai would approve of – is more education for women. This is an explicit part of the UN Development Programme. It gives women more autonomy and allows them to become decision-makers in their communities. Educated women can also participate in the wage economy – some farms now depend on their income as teachers or health professionals, and there are also several handicraft enterprises for women.

You may think this has nothing to do with climate change. However, research by the Brookings Institution showed that the number of years of schooling that girls receive in any given country is strongly correlated with the country's ability to tackle climate change (Murphy, 2018).

Afghan women, and younger Afghans, now know about climate change. They're no longer just victims of forces beyond their knowledge or control; they're demonstrating, they're active, they're demanding an end to pollution, they're bringing change.

Life is still difficult, particularly with the US army now pulling out . . . and it's not Afghanistan which is fueling climate change, after all. But there are small seeds of hope that Afghanistan could reclaim its pristine beauty and become a fertile country once again.

11

CALIFORNIA:
FIRE AS FRIEND, FIRE AS FOE

T HE PEOPLE ARE RUNNING. DEER are running. Garter snakes burrow deep under tree roots. Elk hunker down in the stream, only antlers and nostrils above water. Insect brothers hide in logs or scuttle for safety.

The people are burning. Resins in pine sap spit, flames soar. Willow's leaves wilt as the fire comes closer. Grasses blaze. Fir splits. Oak sunders.

Fire as tall as a tree.

Grandfather in the sky, protect our people.

* * * * *

I'm seeing a wall of fire, so bright my eyes hurt, so hot my skin feels as if I'm burning.

It's been going for days. You see the smoke, you think it's a volcano, you can see it miles off.

It keeps going. It moves. It runs. It jumps. It goes 40 miles an hour, faster than I can run. And I'm up against it with a hose. I'm not going to win.

~ Based on photo by Josh Edelson ~

Used to be, the wind dropped, the fire would go out. Ran out of fuel eventually. Now, the fire's got its own life. It just keeps going.

Can't see properly. Can't breathe properly. But the Fire Department doesn't give up easily.

* * * * *

Joe opens the envelope. He's just finished breakfast on the balcony of his architect-designed house, with a view of the pine-woods and the far horizon. His dream house. He and Marina have worked hard for it. He unfolds the letter, takes a look at it, and says "Holy . . . "

"Now, now, Joe," says his wife. "None of your language."

He passes the letter to her.

"What the . . . "

"Now, now," he says. "None of your language."

Their home insurance has just tripled (Riquelmy, 2018).

* * * * *

If you don't trust the climate scientists, trust the insurance companies. They rarely get their risk estimation wrong. The state insurance commissioner thinks Californian homes could become uninsurable unless something is done about the wildfires.

Joe and Marina are lucky that they can still get insurance from their current insurer, at any price. Insurers dropped their cover on over 42,000 Californian homes in the Sierra Foothills in 2019, almost twice the amount of the year before (Kasler, 2021).

* * * * *

This is the impact of Californian wildfires from different individual perspectives. The high-level impact is equally worrying.

In 2020, 4.3 million acres burned. Even by October, the year's wildfires had broken all records (McKeever, 2020) with over 10,000 structures destroyed and hundreds of thousands of people evacuated – some for a couple of days, others for longer. (Some went home; others had no home to go back to.) One fire chief was working on a wildfire when he realized his own house was going to burn, and he had no resources left to save it; he'd already sent all his engines into action.

Napa Valley and Sonoma County saw huge swathes of vineyards destroyed, as well as cellars housing vast amounts of wine. That's a huge slice of the local economy and it's suffered badly.

Bernie Krause, a sound recordist, had his entire archive of reel-to-reel tape – a life's work – destroyed in a few hours as his house burned in the Sonoma County fires. (Fortunately, he had a single backup, in France.) Ironically, much of his work in nature recording shows how climate change and human intervention have destroyed whole habitats – something that's now happening in California too (Beeler, 2018).

The news shows terrible video footage night after night after night. Mountains ablaze. Fires swamping neat suburban streets where every neat white-painted house has a stripy lawn in front. Families desperately trying to get to safety, driving through air so full of smoke, they can hardly see the road. (Sometimes you can't see the fire, only the smoke.) Orange skies. Airplanes dropping red flame retardant, which never seems to make any impact on the flaming hills below.

What's really scary is that this is no longer "out in the wild" – firefighters can look down from the flaming ridges and see the twinkling grid of the Los Angeles streetlights. It's getting very, very close.

What we don't see on the news so much is the impact on the fauna that lives in the California forests. Animals have a choice: hide or run. Those that don't, won't survive. And those that run have less habitat to choose from. A burned-out bear could end up foraging trash cans in the suburbs.

The narrow-headed garter snake – completely harmless to humans, by the way – is now running out of habitat. The San Francisco garter snake is already on the endangered list, and now other garter snakes are threatened by the fires. They may burrow into holes or under rocks and survive, but once the fire is over, they have nothing to eat. Where ash covers the water, it kills the fish that the snakes eat (Northern Arizona University, 2014).

Deer may manage to stay ahead of the fireline, but later end up trying to find food where everything has been burned or covered in ash.

The fire isn't an unalloyed disaster, though. A few creatures benefit. Predators can do well. Bobcats know where the squirrels and other rodents are going to run; they might even take on a mule deer. Birds of prey will be on the outlook for small mammals, snakes, or smaller birds. They'll do well for a while; though if too many of their prey die in the fire, they could end up hungry as well.

Some plants need fire too. The jack pine won't release its seed unless there is a fire to trigger the process. And the smaller undergrowth plants soon reseed and take advantage of the opportunity to grow without too much competition, or too much shade. Fire can bring fertility as well as destruction.

One forest fire = 35 million cars' worth of pollution

What is really worrying, though, is not so much the short-term impact of the wildfires, but their long-term impact – in particular, on the quality of the air we breathe. The Napa and Sonoma fires produced 10,000 tons of particulate matter (PM) 2.5 (tiny particles with a size of less than 2.5 microns) in two days. That's roughly the same output as California's 35 million vehicles in a year (Spillman, 2017). (Yet we think, "trees good, cars bad." It ain't necessarily so!) PM 2.5 isn't just the main cause of haze in the US: it's also linked to respiratory problems and can worsen existing heart disease.

Where forest fires have encroached on settled areas, and burn down buildings or incinerate cars, they create toxic ash that might contain pesticides, plastics, and even asbestos, arsenic, lead, and other potentially harmful minerals (Franz, 2017). No wonder firefighters who once used to wear cloth masks or simple barrier masks are now using respirator masks as protection.

We don't know exactly what the long-term impact will be. But it won't be good.

* * * * *

Older people remember that the 1960s and 1970s weren't like this. There were practically no fires. The Forest Service made sure of that.

And now, it seems to be fire after fire after fire. Fifteen of California's 20 largest recorded fires have happened in just the last two decades. Every summer now, fires burn eight times the area that they did 50 years ago (Borunda, 2019).

Why the change?

First, longer, drier summers. California has become warmer by 3°F (1.7°C) over the course of the last century, one degree more than the global average (Borunda, 2019). Those summers turn pine trees in particular into nature's firelighters – full of inflammable resin, and all dried up and ready to burn. But any drought-stressed or dead tree is fuel for a fire – and California's full of them; the US Forest Service believes that 80 percent of the fuel for 2020's Creek Fire was made up of trees that had already been killed by bark beetle infestation (Wigglesworth, 2021).

Hotter air draws the water out of those plants that survive – and many don't. So dry undergrowth, grass, and brush are primed to carry the flame.

The weather is more variable than it used to be, too. Mostly, in the fire season, the wind comes from the east, plunging down from the ridge of the Sierras into the valleys on the other side. The Santa Ana winds of autumn can gust at 70–80 miles an hour, and the air they bring with them is not just hot but also very dry. If a fire gets started when these winds are blowing, it will run with the wind – downhill, towards the cities and the coast. (Normally, a fire will never go downhill – a fact that controlled burning and windbreaks rely on.)

These strong winds are getting stronger, more frequent, and sometimes occur at other times of the year; so it's difficult to predict whether a fire will be limited, or whether the winds could take control of it.

And the fire season has become longer – 75 days longer than a few decades ago (Borunda, 2019). That's because there's less snow in the mountains, so it melts earlier, and there's less autumn rain, coming later and later.

Each big fire makes the problem a little worse. High-intensity fires can ruin the soil – they bake it, burn the organic matter that ought to rot down, and cover it with ash. Even mycelium, the underground part of fungi buried deep beneath feet of dirt, and crucial to a forest ecosystem, can be burned out.

Higher temperatures, poorer soil, and less water make it more difficult for the forest to regenerate. In nearly a third of the fire sites that Colorado State University studied, no seedlings were pushing up to replace the burned trees (Stevens-Rumann *et al.*, 2018) – will the land turn into grassy, shrubland chaparral? And if it does, that won't solve the problem; the chaparral is even more flammable than the forest.

People in the wrong places

So far, changes in the climate explain a lot about the fires. Also, lightning strikes, a natural event, accounted for four of California's largest ever six fires (Wigglesworth, 2020). But people in the wrong places – and doing the wrong things – don't help. One fire in 2020 was set off by a couple using blue-for-a-boy fireworks at their baby gender-reveal event (Moon and Silverman, 2020). That was the El Dorado Fire, which affected 10,574 acres (16.5 square miles). A previous gender-reveal party in Arizona in 2017, ignited a wildfire that caused over $8 million in damage and spread over 47,000 acres (73 square miles).

Research from the Conservation Biology Institute in Corvallis, Oregon, shows that while climate change is an important factor in forest fires, it's not the only one. Human presence is found to correlate significantly with wildfires (Schmidt, 2018). It can also make an existing wildfire worse; houses and cars can burst into flame from the heat.

Besides, flying embers can start a new blaze, so residents have often resisted controlled burning operations. People are afraid of fire in general, even though, unlike a wildfire, controlled burns don't reach such high temperatures or create such a mass of embers (Johnson, 2019). (We'll talk more about controlled burning later – and why people should welcome it rather than fear it.)

Where humans live, they need to be supplied with utilities. That means power lines, with obvious potential for ignition. There have been several fires recently due to PG&E Corp (Pacific Gas and Electric) not trimming and removing trees near its high-voltage power lines, and it has now been put on notice by the state observer that it needs to sharpen up. Liability for a series of fires, including the 2018 Camp Fire – the deadliest and most expensive global natural disaster –, was the reason for PG&E's bankruptcy, from which it only just emerged. It clearly still hasn't learned its lesson.

Humans have changed the forest, too, by altering native vegetation. Plantations have closely spaced trees; when they grow to maturity, their branches interlock, so once one tree is on fire, it will set fire to all the trees around it. Typical plantations are also monocultures – not only because only one species is planted, but also because the high density prevents undergrowth from getting established.

Decades of pine needles and debris can keep rainwater from reaching the roots of the trees. The trees begin to dry out – and so the forest makes its own fuel.

Those pine needles burn pretty well, too.

To find a solution, we need to rethink some of our "green" attitudes. More trees is not always better; more space with fewer trees might be more sustainable.

And "trees good, fire bad" – our immediate reaction when we see the news stories of raging wildfires – might need to be rethought, too. Maybe fire, too, can be good.

* * * * *

Not too much fire, but the wrong kind of fire

California's wildfires present a problem for environmental campaigners. It's difficult to get the message right because there are many different issues involved. The causal chain "climate change, getting hotter, fire" is simple – but it's also simplistic.

Fire is a natural part of the Californian forest ecosystem. We should be talking about what *kind* of fire, not simply the fact that fires are happening. And this needs to be an in-depth discussion; otherwise, for instance, you get timber companies claiming that extensive clearcutting – felling all the trees on a given acreage – is the best way to protect the forest (Bevington, 2018). It isn't.

As for burning logged wood as biomass, that's the worst example of faux sustainability.

Climate activists who simply use the negative aspects of wildfires as a bogeyman, without understanding the place of fire in the forest ecosystem, could hurt their cause. People could start to believe there is too much fire in the forest. There isn't. There's the wrong kind of fire.

Elderly Californians remember how there were no fires in the '50s or '60s. But they're not old enough to remember the forest in the 19th century. There were a lot more forest fires a century ago. Fires were suppressed mechanically as standardized industrial management became a model that we applied to nature as well as the manufacture of automobiles. In fact, that created a system where there was *not enough* fire to allow the forest, as it were, to "let off steam" – some scientists think forests need five times more burn than they were actually getting.

In a natural forest, or a controlled-burn forest, not all fires will be high intensity. There will be a mix of high-, medium-, and low-intensity fires, producing different effects and distinct styles of vegetation. In a high-density forest, it's all high-intensity fire, and that's damaging.

Even severe, high-intensity wildfires have their place in the ecosystem. "Snag forest", with bare soil and standing dead trees, has a life cycle that takes it from being a marvelous carpet of wild-flowers in the first couple of years to a wild-seeded nursery of young trees by the fifth or sixth year after the fire. And all those trees are sequestering more carbon as they grow (Bevington, 2018). But the forest also needs the low-intensity type of fire that just burns away undergrowth and debris – cleaning up the forest.

Of course, another thing that's changed dramatically in the last century is the number of people living in these areas. Cities have grown closer and closer to the wildlands and more and more people are living close to or even in the woods. More wildfires mean those people, and the authorities responsible for them, must adapt. They have to learn how to evacuate quickly and efficiently, and part of that is about getting people to trust the call to get out.

Previously, residents might get 10 or 20 minutes to leave, and they'd hang on until the last moment, trying to remember if they'd left anything they needed, or calling the cat. Now, authorities tend to act earlier, with a larger evacuation zone. That gives residents longer to organize themselves and get out. They may get a cellphone alert; in some places, police or fire officers will knock on doors to ensure everyone has got the word.

In Healdsburg City, the decision to evacuate was made at 10 in the morning, and the whole town of 11,500 residents was evacuated by four in the afternoon.

Better planning comes from experience, too. The first time Sutter Santa Rosa Regional Hospital had to evacuate, it was in a rush – medical staff could already see the fire and smoke as they got the last patient out, though fortunately, the fire didn't get much closer. The next time, in 2019, it took just 12 hours to evacuate the hospital, moving all 86 patients – including some in intensive care – by ambulance or helicopter (Kelly, 2019).

But evacuating your home every couple of years – or even more often – is no way to live. Maybe we need not build houses and communities out in the wild. Or perhaps, we could build better homes. "Home hardening", suggested by state insurance commissioner, Ricardo Lara, as a way of ensuring properties in fire-prone areas are still insurable, could be one way to go.

For houses built since 2008 in CalFire-designated fire-prone areas, home hardening has been mandated by the building code. It involves installing fire-resilient roofing for example, instead of wooden shingles. It requires homes to be buffered by "defensible space", with shrubs well trimmed and no trees close to the house. There are even precise specifications for the mesh on attic vents, to keep flying embers out of the roof space.

And it works – to an extent. A study of houses that survived the Camp Fire showed that while almost 80 percent of older houses were destroyed, just over half the houses built after the 2008 building code were saved. But so far there is no retrofit scheme – and most of the housing in these counties was built before the 2007–8 credit crunch and ensuing recession, so it's not up to the 2008 code standard. It can cost tens of thousands of dollars to replace a roof; so most owners of older homes aren't doing it.

Building better communities can help, too. Not clustering houses so tightly, and making firebreaks as part of the plan, stops fire from spreading from house to house or from one part of a community to another (Kasler, 2019).

Fight fire with fire

Oddly enough, it turns out rock band Metallica has the right idea – one of the best ways to fight fire is, indeed, with fire.

The Native Americans were doing it a long time before the US Forest Service arrived and changed everything with its "10 am policy" – to put out all fires by 10 the next morning. They engaged in cultural burning – controlled burns as a part of forest management, to clear underbrush and encourage new growth. The forest was their home, their main resource for food, craft materials, and housing, and their sacred place. They looked after it.

For a century, cultural burning was banned. Western science didn't understand it was a natural part of the ecosystem. Now, Indigenous tribes, including the North Fork Mono, are working with various agencies to revive the custom.

To the Native American, the entire ecosystem is managed by the people. The forest has to be burned so that the deer can learn what woodsmoke smells like and where to run. It has to be burned so that young, straight branches will grow for basket-making. It has to be burned so berry bushes and wild herbs will grow for medicine. It has to be burned so the tall tree will learn from the charring to grow its bark strong and resistant.

Cultural burning results in a relatively open woodland; you can see through the trees. You can walk easily through the land – there are no fallen logs uncleared, and the undergrowth isn't so thick and thorny that you need to push your way through it. It increases biodiversity, and it reduces the fuel content of the forest from 60 tons an acre on uncleared properties, to just 10 tons on those that have been control-burned (Johnson, 2019).

Local communities are now adopting controlled-burn techniques, often through residents' associations. It's a communal effort, rather than each household looking after itself. In a way, you could even say some of these communities are learning from the Native Americans how to bond with each other, as well as how to channel the creative properties of fire.

Fire moves uphill, since warm air rises, so you start a controlled burn just below a ridge, on a windless day. A drip can full of gasoline and diesel dribbles out a line of flame to start the fire, a narrow strip that moves to the top of the ridge, then dies. You walk downhill a little, and you burn the next strip until the fire reaches the scar of the first burn, and in its turn, dies out.

You burn up all the old leaf mold, all the pine needles, all last year's leaves lying dry and rustling on the ground. You get rid of the excess fuel. When all you have is ash, you rake it in, so the ash is not left lying for the wind to pick up and scatter, but turned into the soil to be mixed and slowly absorbed (Sommer, 2020).

There are still a few issues to be addressed, for instance, insurance considerations (Kasler and Sabalow, 2020). If a controlled burn, for whatever reason, gets out of hand, who is liable? Senator Bill Dodd is now introducing a law (Senate Bill 332) that would restrict the liability of certified-burn bosses; as long as they haven't been grossly negligent, they shouldn't be liable. But it hasn't been passed yet.

* * * * *

Native people are now at the forefront of the controlled-burning movement, though many, because of the way the US government has treated them in the past, are reticent about working with government organizations. (Weaver Tima Lotah Link notes the irony of signing into a nature reserve that her tribe used to own.)

And there is something inspiring about the way Native Americans live with their environment. For instance, a basket has a prayer woven into the lid to protect the plants from which the materials were taken; and people sing in the forest as they burn, or as they gather weaving or medicine plants. It feels like a dance between man and nature.

Ron W. Goode, North Fork Mono Tribal Chairman, is one of the elders teaching a new generation how cultural burning protects nature. He says the forest is his people's home, and you keep your home clean. An unburned forest is like a trashy backyard. It needs a cleanout.

"Our forest is trashy," he says. "You cannot leave it like that. You have to have fire in order to have rejuvenation."

12

KENYA:
The Old and the New

A BOVE YOUR LITTLE GROUP, TALL trees reach to the sky, their leafy canopy providing shade. Dry leaves scatter the forest floor. You're walking single file down a path towards the village. Bees are buzzing.

The glossy leaves of banana plants reflect the sunlight. It's baking out on the plain, but here under the trees, it's cooler. There's even a breeze blowing. Birds call from high in one of the trees. In a clearing, you see neat beds of bright-green foliage.

Is this a forest? Or is it a farm? You're not sure. Could it be both?

Bright red fruits hang from vines. You think about reaching for one but wonder if it's edible or poisonous. You look at your guide.

"It's fine," she says.

It tastes strangely like a tomato.

* * * * *

In Kenya, agriculture is still the main sector of the economy, and yet food insecurity is rife. The UN Food and Agriculture Organization says a fifth of the entire Kenyan population is food insecure – that means people don't have consistent access to enough food for a healthy and active life (McDonnell, 2016).

One issue is that Kenya has twice the global average population growth, so there are always more mouths to feed. But that's not all.

Kenya has a big problem because much of its agricultural methods – some dating from colonial times – no longer match its climate. Since the 1980s, temperatures have risen 2.7°F (1.5°C). The dry season is 40 percent longer, and rainfall has become unpredictable and unevenly distributed, so farmers can't rely on rain for their crops. It's extreme; a mix of drought and downpours.

Arable land that used to be viable is now difficult to farm; the soil has become impoverished, and there's not enough water. Yet home-grown cereals aren't competitive enough with cheap grain imports for farmers to make a reasonable return if they fertilize and irrigate. It's a no-win situation.

But there is a solution – agroforestry.

From big prairies to multi-cropping woodlands

Modern western-style agriculture isn't sustainable here. Big fields, single crops, no trees or shade anywhere – this isn't a model that works. Focusing on yield per acre and short-term gains isn't sustainable in the long term. Mechanical irrigation removes loose topsoil and contributes to soil erosion, and too much plowing encourages the release of soil carbon. Too much artificial fertilizer and excessive use of pesticides have destroyed biodiversity. The damage spreads way beyond the fields; nitrous fertilizer can contaminate the groundwater, and can lead to phytoplankton and toxic algal blooms taking over lakes and pools.

"Modern" agriculture methods are expensive. They require huge scale – range farming creates a monoculture, with its extensive fields of maize and wheat. That has driven away some species – the sacred ibis, once often seen, has gone, and local wetlands are declining (Anwar, 2020).

Smaller farmers can't compete. Two-thirds of sub-Saharan farmers don't use fertilizer, and only 3 percent are able to irrigate – these things cost too much for a near-subsistence model, and equipment is difficult to come by (McDonnell, 2019).

So, if the model doesn't work, let's change the model. By combining trees, shrubs, and ground crops, agroforestry creates a diverse and self-sustaining form of farming. You'll find a field of cabbages growing in neat rows between high palms, or a grove of fruit-bearing trees with hives attached to their trunks and bees busily gathering pollen.

Trees can include fruit-bearing varieties like African cherry, Cordia africana – a native species – as well as calliandra, avocado, and macadamia. Enset, "wild" or Ethiopian banana, is indigenous and is grown along streams, where its roots stabilize the banks. Its fruit is sweet, and the leaves make good cattle fodder – contributing to yet another separate income stream. Other alternative sources of income include firewood, lumber, and medicinal plants.

Agroforestry is low cost and sustainable. It's also a return to a traditional style of smallholding and in many ways is similar to the permaculture approach to cropping. Just looking at an agroforestry farm, you can see the contrast with the highly mechanized, irrigated style of agriculture. A bonus for the farmers is that agroforestry uses native strains of seeds, rather than the expensive commercial seed stock used by "modern" farmers, which are often infertile hybrids. That lets farmers save seed for the next season instead of having to buy more.

Agroforestry is a complete system, just like a natural forest. The deep tree roots prevent soil erosion. The soil can store more water when there is sporadic rainfall. And the trees' shade protects the crops from too much sunlight and excessive warmth. Where leaves are used for livestock fodder, in a virtuous circle, the manure goes back onto the land to re-fertilize it. Leguminous trees, in the pea family, like acacia, pull nitrogen from the air and release it into the ground, dramatically increasing crop yields. And tree roots bring up water from as deep as 20 meters below the surface, helping to keep smaller crops watered without using artificial irrigation (Ortolani, 2017).

There's a further benefit for farmers. The "modern" system depends on just two planting seasons – if anything goes wrong, that's a whole harvest written off. But with agroforestry, farmers are producing different crops all year round, giving them a steadier income with less risk of a disastrous harvest (Ortolani, 2017).

They do need to remember to keep their cattle in a shed and fence off the goats, though. Otherwise, those temptingly sweet cabbages could be gone in two shakes of a goat's tail . . .

Farmers have also rediscovered many indigenous plants and started to explore their uses. Using indigenous rather than foreign species helps optimize the crop for the land, and saves species that have been disappearing, threatened by forest encroachment and illegal harvesting, as well as climate change (Njagi, 2018).

Another virtuous circle is that the trees sequester carbon – helping Kenya meet its Paris Agreement goals (Njagi, 2018). Erratic weather has been linked to the loss of tree cover; so again, agroforestry helps address more than one problem. It's a partial solution to the over-harvesting of the forests in previous years. Some farms can support 100 species of trees and shrubs (McDonnell, 2016), increasing biodiversity and providing a habitat for wildlife. Even better, birds and insects attracted to the woods will eat mice, worms, larvae, and other pests, helping farmers avoid the need for harmful pesticides.

(Those birds you heard singing? They're a farmer's friends; they eat insects. Some farmers even hang up nesting pots for them.)

Though the level of biodiversity will never be quite as high as in a natural forest, a forest farm is many times more biodiverse than the "modern" field of wheat. And though it may look like a wilderness or a smallholding, an agroforestry farm can be highly productive (Ortolani, 2017). This is not subsistence farming; this is abundance farming.

The problem is that this kind of farming doesn't happen overnight. It takes time to establish. Some trees take 10 years to reach a reasonable size, and like children, need care and attention through those early years.

It also needs finance. Most Kenyan farmers are already living on the edge; they have no spare money to invest. So initial support has to come from NGOs and the government. Some systems give farmers forest land, which they are expected to look after to grow trees as well as crops. Where NGOs have helped farmers get started, it's transformed their farms – and their lives.

Farming to liberate women

But there's another aspect to all this. In much of Africa, it has traditionally been women who were responsible for much of the agriculture. Sometimes, they're left running the farm when the men of the family head for the city to try to get a job delivering cash wages. In some traditions, seeds are the exclusive responsibility of women, a sacred trust (Kamonji, 2019).

But if they wanted to invest in the farm, Kenyan women always had to ask their husbands for money – they may have done all the work, but they didn't control the family budget.

Things are changing. Now, women have access to microlending, women's groups are helping them become more efficient, and they're at the forefront of agricultural innovation. For instance, GROOTS Kenya, a "grassroots community", ensures women are in charge of their destiny and participate directly in all decision-making processes. Good networking also means that once something works on one farm, the idea is quickly taken up elsewhere – even in the slums of Nairobi, and now in Zambia and Zimbabwe as well (Erbentraut, 2015).

For instance, sack farming is one way to beat the drought. Big plastic bags are filled with a mix of soil, animal manure, and pebbles. Vegetables such as kale, beet, and spinach are grown out of holes in the side of the bag, as well as at the top. The manure provides goodness, the pebbles stop the bag from becoming waterlogged, and because the sides are plastic, the moisture stays in the soil instead of evaporating.

Nairobi found another advantage to sack farming: because the vegetables grow vertically, it uses very little land. Even in a cramped housing compound in the city, there's enough room for a small sack farm.

~ Based on photo by Tim Mcdonnell ~

It's not just women's farms that are being changed – it's their entire lives. Women who started out just trying to survive now understand concepts like nitrogen-fixing and its place in soil fertility. Some attend international conferences, like Purity Gacaga who, though she still takes the *matatu* – privately owned minibus – into town to sell her pop-bottles of milk, joined a delegation at the Paris summit, and travels around Kenya to advise other farmers how to adopt agroforestry techniques (McDonnell, 2016).

Other women have been rediscovering indigenous foods such as sorghum, millet, and cowpeas. Unlike maize, wheat, and rice which are grown in massive single fields, these plants can be grown with less intensive techniques. Also, importantly, they are not traded on international markets, so prices are more likely to remain within most Kenyans' budgets. Since they are not grown with hybrid, infertile seeds, the seed can be saved for planting, de-risking agriculture, and making farmers less dependent on international firms and the capital economy (Kamonji, 2019).

Africa: Silicon Rift Valley

Kenya has also been helped by advanced technology. It's easy to think of Africa as living in some kind of pre-internet golden age, but in fact, Kenya has been at the forefront of micropayments since 2016 with its M-Pesa system. Three-quarters of Kenyans have a cellphone or at least access to one, and M-Pesa accounts hold 40 percent of the country's savings (Rosenberg, 2019).

Farmers now get special weather texts on their mobiles, so they can take action if sudden rain or an unusually dry period is forecast. The Meteorological Department's data isn't always user-friendly, so it uses "climate information intermediaries" who translate the information into farmer-friendly terminology – and make sure farmers get the message.

Huruma Women Group is one of those trusted intermediaries, getting the weather forecasts out to farmers in the region. For instance, mango growers need to know if they will have the 26 hours of sunshine needed to dry the fruit in a greenhouse dryer, or if they should concentrate on the faster-drying vegetables instead. Farmers also get predictions of the start of rainy and dry seasons so they can plan ahead, and the Huruma Women Group even adds recommendations on specific varieties to plant, to take best advantage of expected conditions.

Armed with this information, the group is making more money and has been able to give loans to its members to buy water tanks; a larger processing unit is on the way, too (Mbugua, 2016).

Technology means micropayments, and now it means micro-insurance as well. The costs of insuring small farmers' harvests are prohibitive for big insurance companies, but cellphone sign-up and payout using M-Pesa has transformed the business model. A farmer can now insure two bags of seeds, bought for $10, for just $1 – half that if she buys products from the insurer's partners. The insurance system's solar-powered, automated weather stations give it information such as how much rain fell and at what times. If there's a drought in the growing season or rain too late in the season that wrecks the crop, the insurer automatically pays out to all the farmers in that area. No claim or site visit is needed.

The name of this fintech business is Kilimo Salama – "safe farming" in Swahili – and it's part of a sustainable agriculture non-profit foundation. It's growing – even though many farmers who got their fingers burnt in the past regard insurance as unaffordable and insurers as crooked – because farmers have got to know it and understand that it's honest and reliable. They know someone who received compensation when a crop failed last year; they know someone else who told them how low the payments are – word of mouth is doing Kilimo Salama's marketing for it.

Insurance is particularly important because most farmers are financially so close to the line. If they lose one crop, they may not have enough money to pay for supplies for the next planting season (Rosenberg, 2019). Even with agroforestry, some cash input is needed. But with insurance, one bad harvest won't ruin the farm.

Microlending helps farmers pay for new seeds and new crops to diversify their produce further. Microlender One Acre Fund even gives loans in the shape of seed and fertilizer delivered directly to the farm, rather than cash, and bundles the order with insurance against lost crops.

Instead of "Computer says no," it's "Hakuna matata."

* * * * *

"This one we call *ciakaungi*. This one, *mututwa*. This one, *mugoi*."

Tall white spires of seed. One is knobbly, another flocked with fur. One has piebald seeds, white and gray. Another has little spikes between the seeds that stop birds from eating it. Another looks as if a glue stick has been dipped in popcorn.

Then there is a little hand grenade of brown and white seeds, a kind of sorghum.

To the uneducated eye, most of these seeds look similar. But an experienced woman farmer knows the difference. One is bitter. Another is good for porridge for children. Another you can cook like rice.

* * * * *

And this is just the start. Kenya is now seeing a business boom in new farming methods, equipment, and services. Call it Agtech, Agriculture 2.0, whatever – innovation is changing everything.

There's a new generation of young university-educated Kenyans from farming families, who don't all want to move to the city and get finance jobs; they want to make agriculture more efficient. Their entrepreneurial spirit, helped by microlending as well as by a start-up incubator at the Kenya Climate Innovation Centre

(KCIC), Strathmore University in Nairobi, is beginning to make an impact. KCIC takes funds from major international agencies (the kind that 20 years ago were investing in those massive "modern" fields of wheat) and turns them into microfunding.

One of these new businesses is The Bug Picture, which solves two problems – the need for animal fodder and compost, and the locusts which make farmers' lives a misery. Locusts may look like harmless grasshoppers, but a swarm of locusts blackens the air and can eat its way through a cropped field in a matter of hours; they're immensely destructive. The Bug Picture pays farmers for the locusts they collect, then turns them into insect protein. Locusts are as protein rich as meat and can replace expensive imported fish or soy protein. To help farmers collect the bugs, the company makes collapsible traps and vacuum technologies.

The company already raised black soldier flies as animal feed, but it found out locusts also work well. They find locust swarms and show the local community how to harvest the locust – waiting until night, when the lower air temperature immobilizes the insects, then shaking them out of the trees onto cloths spread out below. In a single night, a family can make a month's wages (Kusmer, 2021).

Another company helping farmers in sustainable ways is Safi Organics. Rice chaff, the hard shell of the rice grain, is normally seen as a waste product, but Safi turns it into high-quality, organic fertilizer. The founder – who had the idea for the business when he was selling his own produce and saw the rice chaff being thrown away – has quit his farm and now runs this business full time (McDonnell, 2019).

Solar power is being used for water pumps; natural pesticides are being produced and tailored specifically for the Kenyan market. Caterpillars are infected with a virus, squashed, and made into a liquid which can then be sprayed – Kenya Biologics' modern way of using an ancient traditional method. And since it's marketing to smallholders, which the big foreign firms don't usually find profitable, it has the market to itself.

The Western world still often treats Africa as a third-rate market where you can sell off obsolescent or second-hand products. But in fact, Africa isn't waiting for Western hand-me-downs. Kenya is innovating its way into a whole new agricultural future, using traditional farming techniques and advanced technology to meet basic needs in a sustainable way.

PART III THINKING POINTS

A CROSS THE WORLD, CLIMATE CHANGE is impacting people and places in very different ways. We've already looked at two of these which are easy to understand: rising temperatures and higher sea levels. But climate can change in less expected ways, particularly at the extremes. In India, which already has extremely high temperatures, heat waves are becoming both more frequent and more intense; in California, forest fires have become a regular occurrence instead of an exceptional one, and they've gotten much bigger.

In California, we see again two of the themes we've already examined – Indigenous land stewardship, and changing the way land is managed. Instead of intensive monoculture plantations, more mixed woodland and shrubland together with a pattern of controlled burning can minimize the chances of out-of-control wildfires.

India is a particularly interesting case for any city-dweller because its major cities have hit the limits earlier than other megacities. Preventing urban heat islands making cities literally too hot to live in can involve some basic citizen actions, like painting roofs white, but could also lead to new urban planning – more trees, more wind corridors, and more public places specifically for people to cool down.

Thinking points:

- Has the weather in your area become less predictable over the past decade or so? If so, think about the impact on business. For instance, do local bars' takings fall if there's a week of stormy weather in midsummer? Do sudden freezes or storms make construction work difficult to schedule?

- The tendency to build megacities has been very pronounced over the past century, particularly in Asia. Can it continue? Singapore, for instance, has tried quite hard to "green" itself, but how effective can that be in densely populated areas?

- Afghanistan shows that climate change can make a particularly toxic mix when combined with existing conflict situations. Do you see trouble spots emerging in other parts of the world?

CONCLUSION

M ANY PEOPLE FEEL POWERLESS IN the face of global climate change. They believe in it, they wish they could do something about it, but they just don't feel able to. When you see the huge amount of plastic produced by your local supermarket, or the amount of traffic on the roads, you wonder if taking your home-sewn shopping bag or using your bicycle to get to work is going to make any difference.

But a lot of these stories have been about ordinary people – often, the poorest and most disadvantaged – taking their future in their own hands.

Human life is fragile. In the developed world it's easy to forget how, in some societies, a single bad harvest can plunge an entire community into poverty, or even famine. But while we are a very adaptable species, we can't cope with extreme heat or cold, extreme thirst or hunger. Often, the people most affected by climate change are already the most vulnerable – the poor, women in patriarchal societies, slum dwellers.

And even developed societies are fragile, as the examples of the Netherlands and California show. (World financial markets haven't yet reacted to the fact that Wall Street is only 20 feet (6 meters) above sea level . . . but they will!)

Twenty or thirty years ago, the approach to such problems would have been top-down. The World Bank and the IMF would have decided where to lend money, financed huge projects, and local people would have been told to do as instructed. But over the decades, we've found that such projects don't always work well; sometimes they have negative side effects, or even make things worse. Lasting solutions to climate-change adaptation must be a combination of bottom-up and top-down – making sure the people whose lives are affected have an equal role in decision-making.

While researching this book, I've been struck by the importance of equality. Empowering the powerless or neglected – Indigenous peoples, women, youth, small nations, poor farmers, and slum-dwellers – leads to imaginative and effective solutions. Empowering people at a grass-roots level has been central to helping societies adapt to climate change.

I started my journey by struggling with climate-change information and communication; all the numbers and long-term policies were abstract and unrelatable. So I decided to look at climate change from a local, personal perspective. I began to appreciate how the impact of climate change on a Pacific island is very different from its effect on a high-altitude desert, and how places you wouldn't consider in connection with climate change are facing dramatic changes – like Burgundy with its vineyards.

I realized, as well, that climate change often has surprising links with other issues, like domestic violence or the disenfranchisement of Indigenous peoples. In Afghanistan, the impact of climate change on farmers has helped the Taliban and led to the large-scale growing of opium poppies.

But I also realized something else. The fight against climate change isn't just about saving the planet. Suppose we make the last stop of our journey the International Space Station, float out into space, and look down at the Blue Planet that we call home.

Astronaut William Anders, who was on the Apollo 8 mission in 1968 and took the famous "Earthrise" photo, said:

"The Earth we saw rising over the battered gray lunar surface was small and delicate, a magnificent spot of color in the vast blackness of space. Once distant places appeared inseparably close. Borders that once rendered division vanished. All of humanity appeared joined together on this glorious-but-fragile sphere."

As fragile as it is, climate change is not about saving Earth.

Earth will carry on – with or without human beings. It has suffered five catastrophic mass extinctions in the past 500 million years, caused by factors ranging from super-volcanic eruptions to asteroid strikes. More than 99 percent of all life that has ever lived on Earth is now extinct (Greshko, 2019). The dinosaurs died out almost completely, leaving only birds as their distant successors. Earth endures, and new species have evolved to thrive in the new environment – among which, eventually, arrived a smart little biped called *Homo sapiens*.

Earth will be fine. Time will heal any major destruction that strikes her. Species come and go but life is made to endure on Earth, and it will go on with or without us.

The battle against climate change is about saving us, our children, and their children, and even more urgently, our culture. It's about saving Inuit's native knowledge, the Californian Native Americans' understanding of their world and their religious and craft traditions; about revaluing old skills at the same time as acquiring new ones. It's about empowering women and young people whose voices are often ignored.

It's about saving our homes and our values, so we can continue to live happily on this planet.

And yes, we can save the planet too.

* * * * *

I've done my part now. I've taken you on this journey around our changing world, and I hope you enjoyed it and found some inspiration in the tales I've told.

Can I ask you to do something for me? Just share the stories.

Share some of these stories with your friends and colleagues. Next time you run out of things to say, you can ask, "Do you know that the taste of Burgundy wine has changed in recent years?" or, "Do you know why mangrove swamps are so important?" or, "Why do honey collectors in the Sundarbans wear masks on the back of their heads?"

And then you can talk a bit about climate change, how it's affecting people, and how people, in turn, are adapting and finding ways to save their livelihoods and their communities.

You could go further. You could observe how climate change has impacted your local community – maybe in ways you hadn't thought of and that aren't obvious. And it doesn't matter how old you are, or what you do – whether you're a stay-at-home parent, an engineer, a teacher, a banker, still at high school – you can find a way to contribute. Think creatively – find local solutions and adaptations, however small.

And if you do take action, please share *your* stories with our "Humans of Climate Change" community on Facebook (more information in the next section). Because someone, somewhere, might just need a few words from you to get started on their own project.

If you've enjoyed this book, and you'd like to make sure more people get the message, then as well as talking to your own family and friends, I'd be humbled and grateful if you'd leave a review on Amazon. That way, more people will read the book, and perhaps we can set off a real snowball effect to communicate not only how serious a challenge climate change sets us, but also how magnificently individual human beings are rising to the occasion.

JOIN OUR COMMUNITY

T HE "HUMANS OF CLIMATE CHANGE" community on Facebook is created to take further actions on everything we have talked about in this book, from sharing personal stories of climate change to empowering women, indigenous people, and young people.

Part of the proceeds from selling the book will be used toward building the community. Join us and participate in the weekly or bi-weekly sustainability challenges for a chance to get involved and **win money**.

Follow this link: https://community.humansofclimatechange. org or scan the QR code below for further instruction.

ACKNOWLEDGMENTS AND PERMISSIONS

I WOULD LIKE TO EXPRESS my deepest gratitude to Andrea Kirkby and Debbie Emmitt for their valuable inputs during the writing of this book.

Thank you to Francisco Marcos de Araújo for the book cover design, to Prasad Weerasinghe for the amazing pencil sketches, and to Praditha Kahatapitiya for the interior design, formatting, and typesetting.

I would also like to extend my sincere thanks to the following artists and photographers who have given me permission to use their work:

- Camilla Andersen for the photo of a young Inuit hunter.

- Charles Nambasi for the photo of an African child leaning against the wall.

- Marquinho Mota for the photo of an indigenous Amazonian woman protesting.

- AnnMary Ravuda for the photo of herself planting mangroves.

- Amy Yee for the photo of a Bangladeshi farmer.

- Dieter for the photo of the Afsluitdijk.

- Silvia Alessi for the photo of an Afghanistan man.

- Mai Ai Bing for the photo of an Afghan school girl.

- Josh Edelson for the photo of a firefighter in the middle of a wildfire.

- Tim McDonnell for the photo of a Kenyan woman standing on her farm.

Last but certainly not least, I would like to thank Hi Vu for her support, encouragement, and patience throughout the years.

REFERENCES

Introduction

Gustafson, A., Ballew, M. T., Goldberg, M. H., Cutler, M. J., Rosenthal, S. A., & Leiserowitz, A. (2020). *Personal Stories Can Shift Climate Change Beliefs and Risk Perceptions: The Mediating Role of Emotion.* Communication Reports, 1–15. doi:10.1080/08934215.2020.1799049

Nabi, R. L., Gustafson, A., & Jensen, R. (2018). Framing Climate Change: Exploring the Role of Emotion in Generating Advocacy Behavior. Science Communication, 40(4), 442–468. doi:10.1177/1075547018776019

1. The Arctic: The Way of Life

Arvin, Jariel. 22 February 2021. Vox. The latest consequence of climate change: The Arctic is now open for business year-round. https://www.vox.com/22295520/climate-change-shipping-russia-china-arctic

CE Delft. Third IMO GHG Study 2014. December 2014. https://ce.nl/publicaties/third-imo-ghg-study-2014/

CE Delft. Update of Maritime Greenhouse Gas Emission Projections. January 2018. https://cedelft.eu/publications/update-of-maritime-greenhouse-gas-emission-projections-2/

Shea, Neil. National Geographic 8 May 2019 As Arctic Ice Melts, a new Cold War Brews. https://www.nationalgeographic.com/environment/article/new-cold-war-brews-as-arctic-ice-melts

Shea, Neil. National Geographic 15 August, 2019. A thawing Arctic is heating up a new Cold War. https://www.nationalgeographic.com/adventure/article/how-climate-change-is-setting-the-stage-for-the-new-arctic-cold-war-feature

Anon. A Comparative Look at Inuit Lifestyle. https://www.learnalberta.ca/content/ssognc/inuitLifestyle/

Eggertson, Laura. 27 September 2015. Nunavut should declare state of emergency over suicide crisis. https://www.cbc.ca/news/canada/north/nunavut-suicide-1.3245844

Ipellie, A., n.d. Legends of the Inuit people: Poem by Alootook Ipellie. [online] Inuitartofcanada.com. Available at: https://www.inuitartofcanada.com/english/legends/poemtwo.htm

Skura, Elyse. 15 July 2015. Melting sea ice brings danger and opportunity to Canada's Arctic. https://www.dw.com/en/melting-sea-ice-brings-danger-and-opportunity-to-canadas-arctic/a-19401653

Traditional Lifestyles of the Inuit. Oceanwide Expeditions Blog. https://oceanwide-expeditions.com/blog/lifestyles-of-the-inuit

Angeleti, Gabriella. 16 April 2020. The Art Newspaper. Market for Inuit art faces deep freeze after Arctic cruises are put on hold. https://www.theartnewspaper.com/news/inuit-artists-face-economic-losses-amid-the-stop-of-cultural-tourism-and-slowdown-of-the-art-market

Deuling, Meagan. 18 October 2019. New fee for Nunavut tourists intended to benefit Inuit. https://www.cbc.ca/news/canada/north/new-tourism-fee-on-inuit-owned-land-1.5323585

Samenow, J., 2018. Arctic temperatures soar 45 degrees above normal, flooded by extremely mild air on all sides. [online] https://www.washingtonpost.com/news/capital-weather-gang/wp/2018/02/21/arctic-temperatures-soar-45-degrees-above-normal-flooded-by-extremely-mild-air-on-all-sides/

Schreiber, Melody. 23 March 2018. Solving the suicide crisis in the Arctic Circle. Pacific Standard Magazine. https://psmag.com/environment/solving-the-suicide-crisis-in-the-arctic-circle

Yeo, Sophie. 24 September 2018. How traditional food is helping communities in a changing Arctic. Pacific Standard. https://psmag.com/environment/how-traditional-food-is-helping-communities-in-a-changing-arctic

2. Burgundy, France: The History of Wine

Borunda, Alexandra. 30 September 2019. National Geographic. Climate change is changing the flavor of French wine. https://www.nationalgeographic.com/science/article/wine-harvest-dates-earlier-climate-change

Bruntlett, Adam. 27 November 2017. The new Burgundy: coping with climate change. https://blog.bbr.com/2019/11/27/the-new-burgundy-coping-with-climate-change/

French Travel Blog Wine Guide to Burgundy, France. https://francetravelblog.com/wine-guide-to-burgundy-france/

Labbé T. *et al*. The longest homogeneous series of grape harvest dates, Beaune 1354–2018, and its significance for the understanding of past and present climate. Climate of the Past. https://doi.org/10.5194/cp-15-1485-2019

Lawrence, James. 18 June 2020. Wine-Searcher. Climate the Latest Challenge for Burgundy. https://www.wine-searcher.com/m/2020/06/climate-the-latest-challenge-for-burgundy

Lawrence, James. 1 May 2020. Harpers. Climate change creeps up in Burgundy. https://harpers.co.uk/news/fullstory.php/aid/27049/Climate_change_creeps_up_in_Burgundy.html

Tourné, Christophe; Hermel, Cédric. 9 April 2021. Le vignoble de Bourgogne a trinqué avec le gel. France Bleu. https://www.francebleu.fr/infos/agriculture-peche/le-vignoble-de-bourgogne-a-trinque-avec-le-gel-1617954078

Wine Folly. July 8, 2013 - Updated on December 2nd, 2020. A Simple Guide to Burgundy Wine (with Maps). https://winefolly.com/deep-dive/guide-to-burgundy-wine-with-maps/

3. Mali and Ethiopia: From Ally to Enemy

Doucet, Lyse. 22 January 2019. The battle on the frontline of climate change in Mali. BBC News. https://www.bbc.com/news/the-reporters-46921487

Endo, Noriko and Eltahir, Elfatih AB. Increased risk of malaria transmission with warming temperature in the Ethiopian Highlands. 2020 Environ. Res. Lett. 15 054006

Kushner, Jacob. 24 April 2020. A new malaria vaccine sparks hope—but cheaper measures are still useful. National Geographic. https://www.nationalgeographic.com/history/article/new-malaria-vaccine-sparks-hope-cheaper-measures-still-useful

McSweeney, Robert. 18 July 2016. Climate change could curb malaria risk in West Africa by end of century. Carbon Brief. https://www.carbonbrief.org/climate-change-could-curb-malaria-risk-in-west-africa-by-end-of-century

MIT, Eltahir Research Group. Increased risk of malaria transmission with warming temperature in the Ethiopian Highlands. https://www.youtube.com/watch?v=7Lt3rRWXN7E

Noone, Greg. 15 September 2017. The machine and the mosquito: How technology is joining the fight against one of the world's most efficient vectors of disease. https://howwegettonext.com/the-machine-and-the-mosquito-29205f61e811

Oxford, University of. 7 May 2021. Promising malaria vaccine enters final stage of clinical testing in West Africa.

4. Amazon Rainforest: The Last Warriors

Aventure-Life.com. Indigenous People. https://www.adventure-life.com/amazon/articles/indigenous-people

Badia i Dalmases, Francesc. 19 June 2019. Ednei: This is Maró Indigenous Land. openDemocracy. https://www.opendemocracy.net/en/democraciaabierta/ednei-this-is-mar%C3%B3-indigenous-land/

Butler, Tina. 18 October 2005. Pre-Columbian Amazon supported millions of people. Mongabay. https://news.mongabay.com/2005/10/pre-columbian-amazon-supported-millions-of-people/

Butler, Rhett. 1 April 2019. People in the Amazon Rainforest. Mongabay. https://rainforests.mongabay.com/amazon/amazon_people.html

Global Forest Watch. Data retrieved 4 May, 2021. https://www.globalforestwatch.org/dashboards/country/BRA/10/

Griscom, Bronson W. *et al.* National mitigation potential from natural climate solutions in the tropics. 27 January 2020. Phil Trans R. Soc. B 375: 20190126. https://doi.org/10.1098/rstb.2019.0126

Loures, Rosamaria and Sax, Sarah. 21 August 2020. Amazon "'women warriors" show gender equality, forest conservation go hand in hand. Mongabay. https://news.mongabay.com/2020/08/amazon-women-warriors-show-gender-equality-forest-conservation-go-hand-in-hand/

Eaton, Sam. The Amazon's carbon tipping point. 4 October 2018. https://www.pri.org/categories/amazons-carbon-tipping-point

Veit, Peter and Ding, Helen. 7 October, 2016. Protecting Indigenous Land Rights Makes Good Economic Sense. World Resources Institute. https://www.wri.org/insights/protecting-indigenous-land-rights-makes-good-economic-sense

Welch, Craig. 11 March 2021. First study of all Amazon greenhouse gases suggests the damaged forest is now worsening climate change. National Geographic. https://www.nationalgeographic.com/environment/article/amazon-rainforest-now-appears-to-be-contributing-to-climate-change

World Wildlife Organization. The vital links between the Amazon rainforest, global warming and you. https://wwf.panda.org/discover/knowledge_hub/where_we_work/amazon/about_the_amazon/why_amazon_important/

5. Mekong Delta, Vietnam: Rice and Salt

Brown, David. 24 December 2020. Analysis: How Vietnam came to embrace a new vision of the Mekong Delta's future. Mongabay. Retrieved from https://news.mongabay.com/2020/12/analysis-how-vietnam-came-to-embrace-a-new-vision-of-the-mekong-deltas-future/.

Colet, John. (2002) *Footprint Vietnam*. Bath, UK: Footprint Handbooks.

Grootens, J., 2019. Sea Level Rise in the Mekong Delta. [online] ArcGIS StoryMaps. Available at: https://storymaps.arcgis.com/stories/5db105cdf41f4e-a08154ffd9a7cc7918

Lovgren, Stefan. July 31, 2019. Mekong River at its lowest in 100 years, threatening food supply. National Geographic Magazine. Retrieved from https://www.

nationalgeographic.com/environment/article/mekong-river-lowest-levels-100-years-food-shortages

Mason, Florence and Jealous, Virginia. (2003) *Lonely Planet Vietnam, 7th edition*. Melbourne, Lonely Planet Publications.

Southern Institute for Water Planning: Study on Climate Change Scenarios Assessment for Ca Mau Province, Technical Report. April 2008.

Tatarski, Michael. March 10, 2021. As the Mekong delta washes away, homes and highways are being lost. London, UK: China Dialogue. Retrieved from https://chinadialogue.net/en/energy/as-the-mekong-delta-washes-away-homes-and-highways-are-being-lost/

Perlez, Jane. May 28, 2016. Drought and "Rice First" Policy Imperil Vietnamese Farmers. The New York Times. Retrieved from https://www.nytimes.com/2016/05/29/world/asia/drought-and-rice-first-policy-imperil-vietnamese-farmers.html

Troeh, Eve. January 29, 2015. Delta Blues Part 2: When Life Gives You Saltwater, Make Shrimp Ponds. New Orleans Public Radio. Retrieved from https://www.wwno.org/post/delta-blues-part-2-when-life-gives-you-saltwater-make-shrimp-ponds

Vu, DT: Yamada, T and Ishidaira, H. (2018) Assessing the impact of sea level rise due to climate change on seawater intrusion in Mekong Delta, Vietnam. *Water Science & Technology*, Volume 77, Issue 6. Retrieved from https://iwaponline.com/wst/article/77/6/1632/41155/Assessing-the-impact-of-sea-level-rise-due-to

Wallace, Julia. 23 May 2017. Vietnam's response to climate change? A shrimp and mangrove cocktail. The New Humanitarian. Retrieved from https://www.thenewhumanitarian.org/feature/2017/05/23/vietnam-s-response-climate-change-shrimp-and-mangrove-cocktail?

6. Fiji: Paradise at Risk

Dougan, Martin. BBC. 19 September 2020. Climate change: How global warming & rising sea levels are affecting Fiji. https://www.bbc.co.uk/newsround/53556377

Fullerton, Ken. 10 August 2019. Meet Sivendra, the Pacific's Climate Change Hero Without the Cape. https://www.senseandsustainability.net/2019/08/10/meet-sivendra-pacifics-climate-change-hero-without-the-cape/

Met Office. 2021. Development of tropical cyclones. [online] Available at: https://www.metoffice.gov.uk/weather/learn-about/weather/types-of-weather/hurricanes/development

Narang, Sonia. 8 August 2017. Fiji's Climate Champion Speaks Up for Women in the Wake of Cyclones. The New Humanitarian. https://deeply.thenewhumanitarian.org/womenandgirls/articles/2017/08/08/fijis-climate-champion-speaks-up-for-women-in-the-wake-of-cyclones

Narang, Sonia. 9 January 2018. After Devastating Cyclone, Fiji Farmers Plant For A Changed Climate. NPR. https://www.npr.org/sections/the-salt/2018/01/09/573521139/after-devastating-cyclone-fiji-farmers-plant-for-a-changed-climate

Narang, Sonia. 6 November 2017. Life on the Front Line of Climate Change for Fiji's Women and Girls. The New Humanitarian. https://deeply.thenewhumanitarian.org/womenandgirls/articles/2017/11/06/life-on-the-front-line-of-climate-change-for-fijis-women-and-girls

Neimila, Nanise. 27 February 2020. Youth Takes On Climate Change Activism. https://www.fiji.gov.fj/Media-Centre/News/Feature-Stories/Youth-Takes-On-Climate-Change-Activism

New Zealand Ministry of Foreign Affairs and Trade. Subject to Change. https://www.youtube.com/watch?v=VupDgO-4kC8

Rovoi, Christine. 20 September 2019. Fijian student takes climate fight to the world. RNZ. https://www.rnz.co.nz/international/pacific-news/399206/fijian-student-takes-climate-fight-to-the-world

7. Bangladesh: The Rivers Give and the Rivers Take

Bilak, Alexandra. From island to slum: Bangladesh's quiet displacement crisis. March 2019. International Displacement Monitoring Centre. https://www.internal-displacement.org/expert-opinion/from-island-to-slum-bangladeshs-quiet-displacement-crisis

Blue   Gold   Bangladesh.   http://www.bluegoldbd.org/what-we-do/about-blue-gold/

Daniels, William, Keefe, Alexa and Jillani, Jehan. National Geographic. Path of Persecution.   https://www.nationalgeographic.com/photography/graphics/rohingya-refugees-bangladesh-myanmar

Gwin, Peter. May 2014. The Ship-Breakers. National Geographic. https://www.nationalgeographic.com/magazine/article/The-Ship-Breakers

Heitzman, James and Worden, Robert, editors. Bangladesh: A Country Study. Washington: GPO for the Library of Congress, 1989. River Systems. http://countrystudies.us/bangladesh/25.htm

Jordan, Joanne. 03 Feb 2017. Video: the lived experience of climate change; Bangladesh.   https://www.acclimatise.uk.com/2017/02/03/video-the-lived-experience-of-climate-change-bangladesh/

McDonnell, Tim. 24 January 2019. Climate change creates a new migration crisis for Bangladesh. National Geographic. https://www.nationalgeographic.com/environment/article/climate-change-drives-migration-crisis-in-bangladesh-from-dhaka-sundabans

Marsh Sarah. 8 Jan 2020. The Guardian. On the frontline of the climate emergency, Bangladesh adapts. https://www.theguardian.com/world/2020/jan/08/on-the-frontline-of-the-climate-emergency-bangladesh-adapts

Morrison, Dan. 21 April 2014. Fighting back a rising tide. Politico magazine. https://www.politico.com/magazine/story/2014/04/bangladesh-floods-105884/

Samuel, Sigal. 18 August 2019. Vox. This country gave all its rivers their own legal rights. https://www.vox.com/future-perfect/2019/8/18/20803956/bangladesh-rivers-legal-personhood-rights-nature

Schwartzstein, Peter. July 2019. National Geographic. This vanishing forest protects the coasts—and lives—of two countries https://www.nationalgeographic.com/magazine/article/sundarbans-mangrove-forest-in-bangladesh-india-threatened-by-rising-waters-illegal-logging

Sunder, Kalpana. 11 September 2020. BBC. The remarkable floating gardens of Bangladesh. https://www.bbc.com/future/article/20200910-the-remarkable-floating-gardens-of-bangladesh

Thomson, Derek. France 24. 09 January 2017. Silt and sand: the river islands of Northern Bangladesh. https://www.france24.com/en/observers-direct/20161203-silt-sand-river-islands-northern-bangladesh

Yee, Amy. New York Times. 18 November 2014. The floating gardens of Bangladesh. https://www.nytimes.com/2014/11/19/business/energy-environment/bangladesh-farming-on-water-to-prevent-effect-of-rising-waters.html

Yee, Amy. 30 June 2013. "Floating schools" bring classrooms to stranded students. New York Times. https://www.nytimes.com/2013/07/01/world/asia/floating-schools-in-bangladesh.html

8. The Netherlands: Living with the Water

Al, Stefan. 24 March 2020. Why isn't the Netherlands underwater? https://www.youtube.com/watch?v=25LW_PG2ZuI

Baurick, Tristan. 9 March 2020. The Dutch are giving rising rivers more room. Should we follow suit? New Orleans Advocate. https://www.nola.com/news/environment/water_ways/article_2dca0db4-5e56-11ea-9452-e3ac6e96c114.html

Kimmelman, Michael. 15 May 2017. The Dutch Have Solutions to Rising Seas. The World Is Watching. The New York Times. https://www.nytimes.com/interactive/2017/06/15/world/europe/climate-change-rotterdam.html

Mostert, Eric. Water and national identity in the Netherlands; the history of an idea. Water History, 12, 311-329. 2020. https://link.springer.com/article/10.1007/s12685-020-00263-3#Sec9

Viviano, Frank. September 2017. This tiny country feeds the world. National Geographic. https://www.nationalgeographic.com/magazine/article/holland-agriculture-sustainable-farming

9. India: The Invisible Enemy

Berardelli, Jeff. Heat waves and climate change: Is there a connection? 25 June 2019. Yale Climate Connections. https://yaleclimateconnections.org/2019/06/heat-waves-and-climate-change-is-there-a-connection/

Bibek, Bhattacharya. 26 September 2020. Is extreme heat making India unlivable? Mint. https://www.livemint.com/mint-lounge/features/is-extreme-heat-making-india-unlivable-11601034638011.html

Chandra, Shekar. 4 July 2019. Are parts of India becoming too hot for humans? CNN. https://edition.cnn.com/2019/07/03/asia/india-heat-wave-survival-hnk-intl/index.html

Curran, Patrick and Siderius, Christian. Cities, climate change and chronic heat exposure. 2019. https://www.lse.ac.uk/granthaminstitute/publication/cities-climate-change-and-chronic-heat-exposure/

Frayer, Lauren. 2 July 2019. Temps Have Topped 120 In India. How Are They Coping With The Heat Wave? NPR. https://www.npr.org/sections/goatsandsoda/2019/07/02/730378851/temps-have-topped-120-in-india-how-are-they-coping-with-the-heat-wave?t=1616969987188

Raj, Sarath; Paul, Saikat Kumar; Chakraborty, Arun; Kuttippurath, Jayanarayanan. Anthropogenic forcing exacerbating the urban heat islands in India. Journal of Environmental Management, Vol 257, 1 March 2020. https://www.sciencedirect.com/science/article/pii/S0301479719317244?via%3Dihub

Ramirez, Rachel. Heat Waves Kill More People Than Any Other Weather Disaster. These Cities Have A Plan. 12 January 2020. HuffPost. https://www.huffingtonpost.co.uk/entry/heat-waves-cities-climate-change_n_5fad4e44c5b68707d1fcf7e4

Safi, Michael. 31 July 2017. Suicides of nearly 60,000 Indian farmers linked to climate change, study claims. The Guardian. https://www.theguardian.com/environment/2017/jul/31/suicides-of-nearly-60000-indian-farmers-linked-to-climate-change-study-claims

Safi, Michael. 2 June 2018. India slashes heatwave death toll with series of low-cost measures. Guardian. https://www.theguardian.com/world/2018/jun/02/india-heat-wave-deaths-public-health-measures

Woetzel, Jonathan; Pinner, Dickon; Samandari, Hamid; Gupta, Rajat; Engel, Hauke; Krishnan, Mekala; & Powis, Carter. McKinsey Global Institute. 25 November 2020. Will India get too hot to work? https://www.mckinsey.com/business-functions/sustainability/our-insights/will-india-get-too-hot-to-work

10. Afghanistan: Between the Two Fights

Carey, John. 29 June, 2011. Global Warming and the Science of Extreme Weather. Scientific American. https://www.scientificamerican.com/article/global-warming-and-the-science-of-extreme-weather/

Jones, Sophia. 3 February 2020. In Afghanistan, climate change complicates future prospects for peace. https://www.nationalgeographic.com/science/article/afghan-struggles-to-rebuild-climate-change-complicates

Kranz, Michal. 17 photos that prove the country where America has been fighting its longest war is actually one of the most beautiful on earth. 30 December, 2017. Business Insider. https://www.businessinsider.com/afghanistan-is-beautiful-photos-2017-12?r=US&IR=T#amid-afghanistans-mountains-hope-remains-because-as-the-afghan-proverb-reads-after-every-darkness-is-light-17

Mehrdad, Ezzatullah. 1 December 2020. Afghanistan's Biggest Fight: Climate Change. The Diplomat. https://thediplomat.com/2020/12/afghanistans-biggest-fight-climate-change/

Murphy, Beth. 4 July, 2018. Educated Afghan women offer economic resilience in the face of climate change and conflict. PBSO News Hour. https://www.pbs.org/newshour/show/educated-afghan-women-offer-economic-resilience-in-the-face-of-climate-change-and-conflict

Rasmussen, Sune Engel. 28 August 2017. The Guardian. How climate change is a"death sentence" in Afghanistan's highlands. https://www.theguardian.com/world/2017/aug/28/how-climate-change-is-death-sentence-afghanistan-highlands-global-warming

Ratcliffe, Rebecca. 25 March 2019."The country could fall apart': drought and despair in Afghanistan. The Guardian. https://www.theguardian.com/global-development/2019/mar/25/country-could-fall-apart-drought-despair-afghanistan

United Nations Development Program. Climate Change Adaptation Afghanistan. https://www.af.undp.org/content/afghanistan/en/home/projects/CCAP-Afghanistan.html

US Drought Monitor. April 13, 2021. http://www.c2es.org/content/drought-and-climate-change/

11. California: Fire as Friend, Fire as Foe

Beeler, Carolyn. 20 November 2018. After his life's work burned, audio recordist links California fires to the"extinction of whole habitats". The World. https://www.pri.org/stories/2018-11-20/lifes-work-goes-flames-audio-recordist-says-california-fires-are-extinction-whole

Bevington, Douglas. 10 July 2018. Mongabay. Backfire: How misinformation about wildfire harms climate activism (commentary). https://news.mongabay.com/2018/07/backfire-how-misinformation-about-wildfire-harms-climate-activism-commentary/

Borunda, Alejandra. Climate change is contributing to California's fires. 25 October 2019. National Geographic. https://www.nationalgeographic.com/science/article/climate-change-california-power-outage

Cal Fire Dept of Forestry and Fire. n.d. Habitat Sweet Habitat. https://www.fire.ca.gov/programs/resource-management/resource-protection-improvement/landowner-assistance/forest-stewardship/wildlife/

Franz, Julia. 11 November 2017. After wildfires, health risks linger. The World. https://www.pri.org/stories/2017-11-11/after-wildfires-health-risks-linger

Johnson, Nathaniel. 14 January 2019. Friendly Fire: Can neighborhood burn squads save California from the next big wildfire? Grist. https://grist.org/article/my-hometown-is-going-to-burn-heres-how-my-neighbors-are-preparing/

Kasler, Dale. 19 October 2020. Insurance companies abandoning California at a faster rate, as wildfires wreak havoc. Sacramento Bee. https://www.sacbee.com/news/california/fires/article246561448.html

Kasler, Dale. 11 April 2019. "The weakest link": Why your house may burn while your neighbor's survives the next wildfire. Sacramento Bee. https://www.sacbee.com/news/california/fires/article227665284.html

Kasler, Dale and Sabalow, Ryan. 21 April 2021. California "burn bosses" set controlled forest fires. Should they be safe from lawsuits? Sacramento Bee. https://www.sacbee.com/news/california/article250853609.html

Kelly, Heather; Wilson, Scott; Lanzendorfer, Joy. 29 October 2019. As fires rage, California refines an important skill: Evacuating. The Washington Post. https://www.washingtonpost.com/nation/2019/10/28/fires-rage-california-re-fines-an-important-skill-evacuating/

McKeever, Amy. 1 October 2020. Inside California's race to contain its devastating wildfires. National Geographic. https://www.nationalgeographic.com/science/article/inside-california-race-to-contain-devastating-wildfires

Moon, Sarah and Silverman, Hollie. 8 September 2020. A California fire sparked by a gender reveal party has grown to more than 10,000 acres. https://edition.cnn.com/2020/09/08/us/el-dorado-fire-gender-reveal-update-trnd/index.html

National Geographic Staff. 8 September 2020. What do wild animals do in wildfires? https://www.nationalgeographic.com/environment/article/150914-animals-wildlife-wildfires-nation-california-science

Northern Arizona University. July 23 2014. Helping save narrow-headed garter snakes from after-effects of Slide Fire. ScienceDaily. Retrieved April 28, 2021 from www.sciencedaily.com/releases/2014/07/140723110133.htm [NAU]

Riquelmy, Alan. 13 December 2018. Nevada County customers lose home insurance; some see double, triple increases. The Union. https://www.theunion.com/news/local-news/insurance-problems-arent-new-to-nevada-county/

Schmidt, Stephen. 1 January 2018. Forest fire surge may be blamed more by human touch than changing climates. The World. https://www.pri.org/stories/2018-01-01/forest-fire-surge-may-be-blamed-more-human-touch-changing-climates

Sommer, Lauren. 24 August, 2020. To Manage Wildfire, California Looks To What Tribes Have Known All Along. NPR. https://www.npr.

org/2020/08/24/899422710/to-manage-wildfire-california-looks-to-what-tribes-have-known-all-along?t=1617060645718

Spillman, Benjamin. 11 Oct 2017. California fires making more pollution than a year of traffic. Reno Gazette-Journal. https://eu.rgj.com/story/life/outdoors/2017/10/11/california-fires-making-more-pollution-than-year-traffic/756020001/

Stevens-Rumann, C.S., Kemp, K.B., Higuera, P.E., Harvey, B.J., Rother, M.T., Donato, D.C., Morgan, P. and Veblen, T.T. (2018), Evidence for declining forest resilience to wildfires under climate change. Ecol Lett, 21: 243-252. https://doi.org/10.1111/ele.12889

Wigglesworth, Alex. 27 April 2021. California is primed for a severe fire season, but just how bad is anybody's guess. Los Angeles Times. https://www.latimes.com/california/story/2021-04-27/california-is-primed-for-a-severe-2021-fire-season

12. Kenya: The Old and the New

Anwar, Shakeel. 21 September 2020. Modern agriculture and its impact on the environment. Jagran Josh. https://www.jagranjosh.com/general-knowledge/modern-agriculture-and-its-impact-on-the-environment-1518163410-1

Erbentraut, Joseph. 3 August 2015. Kenyan Women Are Quietly Revolutionizing Farming.. And The Government's Noticing. HuffPost. https://www.huffingtonpost.co.uk/entry/kenyan-women-growing-crops_n_55bfad1ce4b0d-4f33a03856a?ri18n=true

Kamonji, Wangũi wa. 5 December 2019. Indigenous women in Kenya rebuild resilience amidst an eco-cultural crisis. OpenGlobalRights. https://www.openglobalrights.org/indigenous-women-in-kenya-rebuild-resil-ience-amidst-an-eco-cultural-crisis/

Kusmer, Anna. 24 March 2021. This start-up turns locust swarms in Kenya into animal feed. National Geographic. https://www.pri.org/stories/2021-03-24/start-turns-locust-swarms-kenya-animal-feed

Mbugua, Sophie. 23 November 2016. Kenya's Women Farmers Get Business Boost From Weather Texts. The New Humanitarian. https://deeply.thenewhumanitarian.org/womenandgirls/articles/2016/11/23/kenyas-women-farmers-get-business-boost-weather-texts

McDonnell, Tim. 23 November 2016. How One Kenyan Farmer Went From "Nothing" to the Envy of the Neighborhood. National Geographic. https://blog.nationalgeographic.org/2016/11/23/this-is-what-a-climate-smart-farm-looks-like/

McDonnell, Tim. 29 December 2019. Could Climate Change Build Big Business in Kenya? National Geographic. https://www.nationalgeographic.com/culture/article/could-climate-change-build-a-business-boom-in-kenya-?

Njagi, David. 31 August 2018. Farmers see promise and profit for agroforestry in southern Kenya. Mongabay. https://news.mongabay.com/2018/08/farmers-see-promise-and-profit-for-agroforestry-in-southern-kenya/

Ortolani, Giovanni. 26 October 2017. Agroforestry: An increasingly popular solution for a hot, hungry world. Mongabay. https://news.mongabay.com/2017/10/agroforestry-an-increasingly-popular-solution-for-a-hot-hungry-world/

Rosenberg, Tina. 9 May 2019. Doing more than praying for rain. Opinionator Blog, New York Times. https://opinionator.blogs.nytimes.com/2011/05/09/doing-more-than-praying-for-rain/

Conclusion

Greshko, M., 2019. Mass extinction facts and information from National Geographic. [online] Science. Available at: https://www.nationalgeographic.com/science/article/mass-extinction?loggedin=true

Made in the USA
Middletown, DE
03 March 2023

26103159R00109